PRACTICAL MATHEMATICS
For Metalworking Trainees

PRACTICAL MATHEMATICS
For Metalworking Trainees

by WILLIAM E. HARDMAN

REVISED EDITION

Revised by:
HARVEY RIBBENS
C. J. (Nick) HYLTON, JR.

NATIONAL TOOLING AND MACHINING ASSOCIATION
TEXTBOOK SERIES
9300 Livingston Road • Fort Washington, Md. 20744 • 301/248-6200

©Copyright 1983

NATIONAL TOOLING AND MACHINING ASSOCIATION

Formerly National Tool, Die and Precision Machining Association

Fort Washington, MD 20744

ALL RIGHTS RESERVED

NTMA Catalog Number 5012

ISBN-0-910399-03-4

Library of Congress
Catalog Card Number
83-62514

FORWARD AND ACKNOWLEDGEMENTS

I developed *Practical Mathematics* for the preapprentice training programs conducted by the National Tooling and Machining Association. The 17,000 students trained in this program since 1964 include thousands of skilled metalworking craftsmen, the backbone of America's machine trades workforce. Many graduates are today shop owners and successful businessmen in their own right.

Math has not changed over the years, but the attitudes of students, the way vocational math is taught and its application in the computer age have undergone revolutionary change. With these factors in mind, I met with industry representatives and educators to develop a format that would best meet the needs they felt were essential. I am indebted to countless persons and organizations for their suggestions and contributions toward helping me accomplish this task.

Harvey Ribbens and C. J. Hylton performed the approved revisions for this edition of *Practical Mathematics*.

Harvey Ribbens is director of the NTMA Training Center, Grand Rapids, Michigan. A journeyman diemaker by trade, Ribbens earned an Ed.S at Michigan State University and has since dedicated 20 years as teacher and administrator in vocational and apprenticeship programs.

C. J. Hylton directs the math department at Herbert Hoover Junior High School in Potomac, Maryland, part of the Montgomery County Maryland School District, nationally recognized for the academic excellence of its graduates. Hylton holds a M.Ed. degree from the University of Maryland with qualifications enhanced by training as a machinist apprentice and applied skills training as a building trades contractor.

William E. Hardman

TABLE OF CONTENTS

EVALUATION EXERCISE
UNIT 1   Evaluation Exercise 3

ARITHMETIC
UNIT 2   Fractions 11
UNIT 3   Addition of Fractions 22
UNIT 4   Subtraction of Fractions 31
UNIT 5   Multiplication of Fractions 39
UNIT 6   Division of Fractions 46
UNIT 7   Decimals 52
UNIT 8   Rounding Off Decimals 58
UNIT 9   Changing Fractions to Decimals 61
UNIT 10   Fraction — Decimal Equivalent Chart 64
UNIT 11   Addition and Subtraction of Decimals 67
UNIT 12   Multiplication of Decimals 72
UNIT 13   Division of Decimals 76
UNIT 14   Decimal Concept Applications 80
UNIT 15   Tolerances 82
UNIT 16   Percent 88
UNIT 17   Signed Numbers 95
UNIT 18   Powers and Roots 105

METRICS
UNIT 19   Metrics 113

ALGEBRA
UNIT 20   Algebra — Addition and Subtraction 127
UNIT 21   Algebra — Symbols and Operations 134
UNIT 22   Algebraic Multiplication 139
UNIT 23   Substituting Numerical Values 146
UNIT 24   Equations — Addition and Subtraction Principles 153
UNIT 25   Equations — Multiplication and Division Principles 157
UNIT 26   Formulas 166

GEOMETRY
UNIT 27   Axioms and Propositions 177
UNIT 28   Angles and Lines 182
UNIT 29   Intersecting Lines and Angles 190
UNIT 30   Relating Lines and Angles to Polygons and Circles 201
UNIT 31   Pythagorean Theorem — Projection of Sides 215
UNIT 32   Circles 232

TRIGONOMETRY
UNIT 33   Introduction to Trigonometry 246
UNIT 34   Functions of Angles 250
UNIT 35   Values of Functions 261
UNIT 36   Right Triangle Solutions 271
UNIT 37   Right Triangle Concept Applications 278
UNIT 38   Law of Sines 306
UNIT 39   Law of Cosines 315

TABLE OF CONTENTS (continued)

Tables of Natural Trigonometric Functions and Squares
and Square Roots ... 325

Glossary ... 357

Index ... 363

EVALUATION EXERCISE

$$? = y + 8$$

UNIT 1

EVALUATION EXERCISE

INTRODUCTION

A skilled machinist as well as a competent machine tool operator must have mastered certain mathematical concepts. This Evaluation Exercise will help identify areas of strength in performing practical mathematics operations. It will also identify areas where the student may need to review or to learn new skills. After evaluating this exercise, your instructor will help you build a strong base that will include the algebra, geometry and trigonometry necessary for you to have a successful career in the metalworking and machining industry.

OBJECTIVE:

After completing this exercise the student will:

- Determine current skill levels in performing arithmetic operations

EVALUATION EXERCISE

ADD:

(1)
```
  64,378
  39,462
   2,249
  38,472
  65,789
  21,106
  50,397
```

(2)
```
1,672,543
  384,657
2,988,785
  363,594
  459,678
```

(3)
```
56,934
89,332
97,968
74,593
33,897
```

SUBTRACT:

(4) 6,843,291
   −6,731,897

(5) 5,983,648
   −2,549,879

(6) 2,874,963
   − 429,389

MULTIPLY:

(7) 385
   x   8

(8) 976
   x 27

(9) 5,283
   x   129

(10) 563,974
    x 59,638

(11) **ADD:**

sixty-three thousandths; _____._____

one hundred ninety-two thousandths; _____._____

two and eighty-four thousandths; _____._____

fourteen and six thousandths; _____._____

two hundred thirty-four ten-thousandths. _____._____

_____._____

(12) **MULTIPLY:**

62.583
x 1.272

(13) **DIVIDE:**

528) 648,937

4

(14) DIVIDE:

3.27) 589

(15) DIVIDE:

.693) 2.5864

(16) ADD:

$$\frac{3}{16}$$
$$\frac{7}{8}$$
$$\frac{2}{32}$$
$$+ \frac{3}{4}$$

(17) ADD:

$$\frac{3}{4}$$
$$\frac{5}{16}$$
$$+ \frac{1}{8}$$

(18) MULTIPLY:

$$\frac{3}{4} \times \frac{5}{16} = \underline{\qquad}$$

(19) FIND:

$$\frac{3}{8} \text{ of } 596 = \underline{\qquad}$$

(20) CHANGE TO IMPROPER FRACTIONS:

$21\frac{3}{4} = \underline{\qquad}$ $5\frac{1}{8} = \underline{\qquad}$ $22\frac{7}{8} = \underline{\qquad}$

(21) Change to decimals:

$\frac{3}{4}$ = ____ $\frac{5}{8}$ = ____ $\frac{7}{8}$ = ____ $\frac{1}{2}$ = ____

$\frac{1}{4}$ = ____ $\frac{1}{8}$ = ____ $\frac{3}{32}$ = ____ $\frac{15}{16}$ = ____

(22) Reduce to lowest terms:

$\frac{18}{32}$ = ____ $\frac{16}{64}$ = ____ $\frac{2}{8}$ = ____

$\frac{12}{64}$ = ____ $\frac{2}{16}$ = ____ $\frac{4}{16}$ = ____

(23) Among the following drill sizes circle the smallest?

$\frac{1}{8}$ $\frac{3}{32}$ $\frac{1}{16}$ $\frac{1}{2}$ $\frac{5}{8}$ $\frac{23}{64}$ $\frac{1}{32}$

(24) Change to fractions of lowest terms:

.375 = ____   .875 = ____   .1875 = ____   .125 = ____

(25) If: 3x + 4x = 21   then   1x = ____

(26) If: a = 2, b = 3, c = 4   then $a^2 + 2b - c$ = ____

(27) ADD

    a)   a + b
            a + b

    b)   3a − 5b + 4c
            3a + 8b − 5c

(28) SUBTRACT

    a)   4y − 8
            6y + 3

    b) $2x^2 - 4x + 4$
            $x^2 - 6x - 4$

(29) **MULTIPLY**

    a) $2x(x - 3)$ b) $(x + y)(2x - y)$

         Ans = _____   Ans = _____

(30)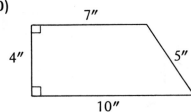

Perimeter = _____"

Area = _____ sq."

(31)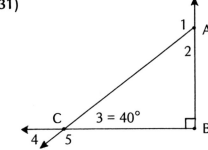

If: $\angle ACB = 40° = \angle 3$

then: $\angle 1$ = _____°

$\angle 2$ = _____°

$\angle 4$ = _____°

$\angle 5$ = _____°

(32) **DETERMINE ANGLES**

$\angle A$ = _____

$\angle B$ = _____

$\angle C$ = _____

$\angle D$ = _____

$\angle E$ = _____

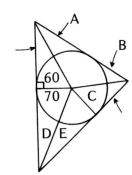

(33) **FIND THE LENGTH OF THE THIRD SIDE**

Ans = _____

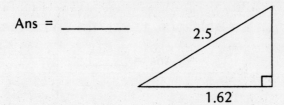

(34) **FIND SIDE A =** _____

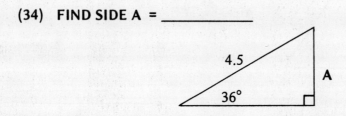

ARITHMETIC

$$\frac{15}{16}$$ 4x7

UNIT 2
FRACTIONS

INTRODUCTION

A machinist's skill depends on the ability to measure. Measurements are seldom accurate enough when expressed in whole units. Most measurements in machining technology involve the use of fractions. They are used for job layout, reading prints, reading a scale and numerous other ways in the shop. Fractions allow us to measure objects with greater precision and are therefore a useful tool. Fractions represent parts of a whole number. Like whole numbers, fractions can be added, subtracted, multiplied and divided. These operations will be developed in Unit 3 through Unit 6.

OBJECTIVES:

After completing this unit the student will be able to:

- define fraction
- identify the set of "shop fractions"
- identify various types of fractions
- express a fraction in different but equal forms

WHAT IS A FRACTION?

A fraction is a part of a whole. Any unit can be divided into a number of equal parts. For example: a standard foot ruler is divided into 12 equal parts. Reference can then be made to one of these parts, two of these parts, three of these parts, etc.

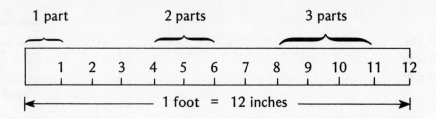

These parts can be related to the total number of parts and expressed with a fraction:

$$\frac{1}{12}, \frac{2}{12}, \frac{3}{12}, \ldots$$

This type of number, the **common fraction**, is made of a **numerator** and a **denominator** written:

$$\frac{n}{d} = \frac{numerator}{denominator} = \frac{1\ part}{12\ parts} = \frac{1\ inch}{12\ inches} = \frac{1''}{12''}$$

NOTE:

A **common fraction** consists of 2 parts:

The **denominator** (bottom number) contains the total number of equal parts in one whole unit.

The **numerator** (top number) contains a number of equal parts which are being related to the total.

DEFINITIONS AND TERMS:

Numbers expressed in this basic form, $\frac{n}{d}$, (or in the form $\frac{1}{12}$, $\frac{3}{8}$, etc.) are called COMMON FRACTIONS.

If the numerator is less than the denominator, such as $\frac{3}{4}$, the common fraction is called a PROPER FRACTION and has a value less than 1 whole unit.

NOTE:
   SHOP FRACTIONS

   The set of proper fractions most frequently used in the shop is:

   $$\frac{1}{2}, \frac{1}{4}, \frac{1}{8}, \frac{1}{16}, \frac{1}{32}, \frac{1}{64}$$

   These fractions are read: one half ($\frac{1}{2}$), one quarter ($\frac{1}{4}$), one eighth ($\frac{1}{8}$), one sixteenth ($\frac{1}{16}$), one thirty-second ($\frac{1}{32}$) and one sixty-fourth ($\frac{1}{64}$).

Symbols: Definition   Important thoughts and ideas

🛑 If the numerator is larger than the denominator, such as $\frac{5}{4}$ or $\frac{3}{2}$, the common fraction is called an **IMPROPER FRACTION**. If the numerator is equal to the denominator the fraction equals 1.

🛑 If the numerator equals "1", these fractions are called **UNIT FRACTIONS**.

The figure at the right is divided into 4 equal parts. 3 have been shaded. The total shaded is $\frac{3}{4}$. What fractional part is unshaded?
Answer: $\frac{1}{4}$

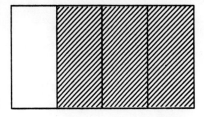

Write a proper fraction to show what part of each figure is shaded. (Relate **shaded parts** to **total parts**.)

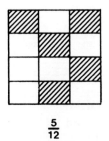 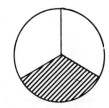 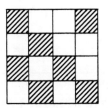 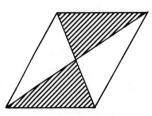

$\frac{5}{12}$ _____   _____   _____

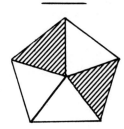 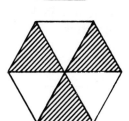  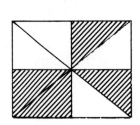

_____   _____   _____   _____

CONSIDER: A square divided into 16 equal sections:

💡 *Step 1* Write a whole number to represent the total number of **shaded** sections. Use this number as the numerator of a fraction.

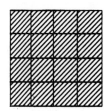

Step 2 Write a whole number to represent the **total** number of sections. Use this number as the denominator of your fraction. Your result is $\frac{16}{16}$. This fraction is another name for the square having a value of **1 whole unit**.

$$\frac{16}{16} = 1$$

13

CONSIDER: 3 equal squares each divided into a number of equal parts.

 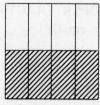 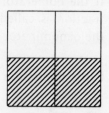

Shaded parts $\frac{8}{16}$ Shaded parts $\frac{4}{8}$ Shaded parts $\frac{2}{4}$

Can you complete the fourth square on the right which has the same shaded area as these three, but represent the shaded area with a simpler fraction? Your fraction is $\frac{1}{2}$.

Therefore $\frac{8}{16}$, $\frac{4}{8}$, $\frac{2}{4}$ and $\frac{1}{2}$ are called **EQUAL FRACTIONS** (or **EQUIVALENT FRACTIONS**) and:

$$\frac{8}{16} = \frac{4}{8} = \frac{2}{4} = \frac{1}{2}$$

The fraction $\frac{1}{2}$ is said to be in **LOWEST TERMS** or **SIMPLEST FORM**. Measurements given with fractions in lowest terms are easier to use and understand. However, frequently when fractions are operated on by adding, subtracting, multiplying or dividing the result is not in lowest terms. One **process** for reducing fractions to their lowest terms is to divide both the numerator and denominator by the **largest** possible number which will go into both numbers evenly (Without a remainder).

EXAMPLES:

$$\frac{8}{16} = \frac{8 \div 8}{16 \div 8} = \frac{1}{2} \quad \text{or}$$

$$\frac{20}{32} = \frac{20 \div 4}{32 \div 4} = \frac{5}{8}$$

CONSIDER:

$$\frac{24}{64} = \underline{}$$

$$\frac{24}{64} = \frac{24 \div 2}{64 \div 2} = \frac{12}{32} \quad \text{, but}$$

$$\frac{12}{32} = \frac{12 \div 2}{32 \div 2} = \frac{6}{16} \quad \text{, but}$$

$$\frac{6}{16} = \frac{6 \div 2}{16 \div 2} = \frac{3}{8}$$

NOW CONSIDER:

$$\frac{24}{64} = \frac{24 \div 8}{64 \div 8} = \frac{3}{8}$$

Which process is best to use and why?

Every **WHOLE NUMBER** can be written as a **FRACTION** by using the number "1" as the denominator and the whole number as a numerator.

EXAMPLE:

$$3 = \frac{3}{1}, \quad 7 = \frac{7}{1}, \quad 24 = \frac{24}{1}$$

A **MIXED NUMBER** is a number made of a whole number added to a fraction, such as:

$$4 + \frac{1}{8} = 4\frac{1}{8} \quad \text{(read as "four and one eighth")}$$

$$3 + \frac{11}{16} = 3\frac{11}{16} \quad \text{(read as "three and eleven sixteenths")}$$

NOTE:

SHOP FRACTIONS APPLIED

Fractions used by machinists are usually in units of $\frac{1}{2}$, $\frac{1}{4}$, $\frac{1}{8}$, $\frac{1}{16}$, $\frac{1}{32}$, $\frac{1}{64}$, or multiples of these: $\frac{3}{4}$, $\frac{7}{8}$, $\frac{9}{16}$, $\frac{5}{32}$, $\frac{11}{64}$ etc. In this text we will use only these common machine **shop fractions** except to illustrate unusual concepts.

RULE FOR SUBSTITUTION:

Numbers which have equal value may be substituted for each other. This is one of the most important concepts in mathematics.

EXAMPLE 1:

(3 + 1) = (5 − 1)   The number "4" can be **substituted** for each
or   4 = 4   of these quantities with the result of 4 = 4.

EXAMPLE 2:

If $\frac{1}{2} = \frac{4}{8}$ and $\frac{1}{2} = \frac{2}{4}$, then by **substitution** $\frac{4}{8} = \frac{2}{4}$ since each is equal to the same number $(\frac{1}{2})$.

IMPROPER FRACTIONS CHANGED TO MIXED NUMBERS

We use **substitution** to show that improper fractions have an exact or equal value in the form of a whole number or a mixed number. The process for changing an **improper fraction** to a **whole number** or a **mixed number** is:

Step 1 Divide the numerator of the fraction by the denominator.

Step 2 If there is a **remainder,** write it over the divisor as a proper fraction.

Step 3 Then reduce the fraction to the lowest terms.

EXAMPLE 1:

$$\frac{24}{8} \rightarrow 8\overline{)24} \quad \text{or} \quad \frac{24}{8} = 3$$
$$\phantom{\frac{24}{8} \rightarrow 8)}\underline{24}$$
$$\phantom{\frac{24}{8} \rightarrow 8)00}0$$

EXAMPLE 2:

$$\frac{19}{8} \rightarrow 8\overline{)19} \quad \text{2 + Remainder 3} \quad \text{or}$$
$$\phantom{\frac{19}{8} \rightarrow 8)}\underline{16}$$
$$\phantom{\frac{19}{8} \rightarrow 8)0}3$$

$$\frac{19}{8} = \frac{8}{8} + \frac{8}{8} + \frac{3}{8} = 2\frac{3}{8}$$

Improper fractions may also be changed to a mixed number using a different form:

EXAMPLE 1:

$$\frac{19}{8} = \frac{16 + 3}{8} = \frac{16}{8} + \frac{3}{8} = 2\frac{3}{8} \quad \left(\text{note: } \frac{16}{8} = 2\right)$$

EXAMPLE 2:

$$\frac{3}{2} = \frac{2 + 1}{2} = \frac{2}{2} + \frac{1}{2} = 1\frac{1}{2} \quad \left(\text{note: } \frac{2}{2} = 1\right)$$

EXAMPLE 3:

$$\frac{35}{8} = \frac{32 + 3}{8} = \frac{32}{8} + \frac{3}{8} = 4\frac{3}{8} \quad \left(\text{note: } \frac{32}{8} = 4\right)$$

MIXED NUMBERS EXPRESSED AS FRACTIONS

You can also write a mixed number as a fraction as follows:

EXAMPLE:
Write $5\frac{3}{4}$ as a fraction.

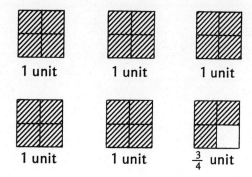

SOLUTION:

Step 1   Multiply the whole number, 5, by the denominator, 4.
$$5 \times 4 = 20$$

Step 2   Add the numerator, 3, to the answer to step 1.
$$3 + 20 = 23$$

Step 3   Write the answer to step 2 as a numerator over the denominator, 4.

Therefore $5\frac{3}{4} = \frac{20 + 3}{4} = \frac{23}{4}$ (shaded region expressed as a fraction)

THE STEEL RULE

The STEEL RULE is the basic measuring instrument used to check dimensions expressed as common fractions. Reading the graduations on the steel rule is often the first practical skill a metalworking student acquires. The illustration shows a rule with the top scale graduated in 64ths of an inch and the bottom scale in 32nds of an inch.

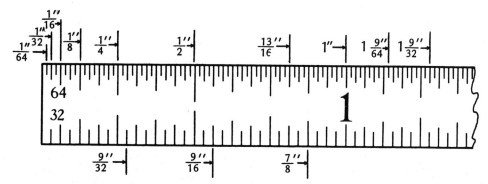

The finest division on the fractional rule is $\frac{1}{64}$th inch. The above scale illustrates the relative size of SHOP FRACTIONS. Each multiple of a shop fraction is marked with an etched line of varying length. The shortest marks indicate $\frac{1}{64}$th inch, the longest represent $\frac{1}{2}$ inch and the whole units are indicated by numbers. Compare the answers from problems (16) and (17) of the CONCEPT APPLICATIONS to the graduations on the rule.

CHECK FOR EQUAL FRACTIONS BY CROSS MULTIPLICATION

When 2 or more fractions have equal value, they are called **equal fractions**

RECALL:

Equal fractions name the same number.

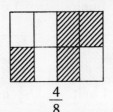 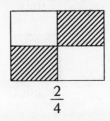 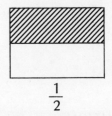

$$\frac{4}{8} \qquad \frac{2}{4} \qquad \frac{1}{2}$$

If two fractions are equal, they have equal **cross products**:

EXAMPLE 1:

Does $\frac{4}{8} = \frac{2}{4}$?   Check the **cross products**

$\frac{4}{8} \times \frac{2}{4}$ means $4 \times 4 = 2 \times 8$
$\qquad\qquad\qquad\qquad\qquad 16 = 16$

Cross products are equal (=), therefore $\frac{4}{8} = \frac{2}{4}$

EXAMPLE 2:

Does $\frac{3}{8} = \frac{1}{2}$?   Check the **cross products**

$\frac{3}{8} \times \frac{1}{2}$ means $3 \times 2 = 8 \times 1$
$\qquad\qquad\qquad\qquad\qquad 6 \neq 8$

Cross products are not equal ($\neq$), therefore $\frac{3}{8} \neq \frac{1}{2}$

The **algebraic form** for **CROSS MULTIPLICATION** is:

$$\frac{a}{b} = \frac{c}{d} \qquad \text{if and only if}$$

or $a \times d = b \times c$

or $ad = bc$

ARITHMETIC FORM AND ALGEBRAIC FORM —
A Simple Difference

You are introduced to mathematics with arithmetic. However, in a short time you will understand and express math in elements of algebra. It is natural and it will be easy. You have already had your first exposure to algebra in the form for CROSS MULTIPLICATION and other forms are given in the next units. We won't elaborate at this time, but the following is a brief explanation of the difference between arithmetic and algebra:

The **elements of arithmetic** are given in the form of **numbers** using whole numbers, fractions, decimals, and percents. These elements may be added, subtracted, multiplied and divided.

The **elements of algebra** are generally shown using **letters or number and letter combinations**. These algebraic elements may also be used in performing the basic operations of addition, subtraction, multiplication and division. They are most important in their use with formulas (formulae). For example, the **area** of a rectangle is determined by multiplying the **length** times the **width**. This process is summarized with a formula made of the elements of algebra: $A = \ell \times w$. To solve this formula, we substitute elements of arithmetic (numbers) to find a specific answer to each specific problem. We will cover the operations with algebraic elements in more detail in later units.

CONCEPT APPLICATIONS

(1) Name three ways fractions are used in the machine shop.

_____ , _____ , _____ .

(2) $\frac{3}{8}$ is a proper or improper fraction? _____

(3) $\frac{11}{8}$ is what type of fraction? _____

(4) $5\frac{1}{4}$ is called a _____ number.

(5) $\frac{16}{16}$ is what type of fraction? _____

(6) A unit fraction must have a numerator which is the number _____ .

(7) $\frac{8}{16}$ and $\frac{4}{8}$ are _____ fractions.

(8) $\frac{5}{32}$ is said to be in lowest terms or in _____ form.

(9) Reduce $\frac{10}{16}$ to lowest terms. $\frac{10}{16}$ = _____

(10) Change $2\frac{7}{8}$ to an improper fraction. $2\frac{7}{8}$ = _____

(11) Change $\frac{19}{4}$ to a mixed number. $\frac{19}{4}$ = _____

(12) Which fractions are most used in the shop? _____

(13) What rule allows you to replace a number with one of equal value?

_____

(14) $\frac{3}{8} = \frac{23}{64}$ True or False _____

(15) Which fraction in problem 14 has the smaller value? _____

(16) Make these fractions equal:

$\boxed{\dfrac{3}{4} = \dfrac{6}{8}}$ $\dfrac{3}{8} = \dfrac{}{32}$ $\dfrac{3}{8} = \dfrac{}{64}$ $\dfrac{1}{16} = \dfrac{}{32}$

$\dfrac{7}{16} = \dfrac{}{32}$ $\dfrac{5}{8} = \dfrac{}{16}$ $\dfrac{1}{2} = \dfrac{}{16}$ $\dfrac{5}{16} = \dfrac{}{64}$

$\dfrac{1}{4} = \dfrac{}{64}$ $\dfrac{1}{8} = \dfrac{}{64}$ $\dfrac{1}{32} = \dfrac{}{64}$ $\dfrac{15}{16} = \dfrac{}{64}$

(17) Write each fraction in lowest terms:

$\boxed{\dfrac{8}{16} = \dfrac{1}{2}}$ $\dfrac{12}{32} =$ $\dfrac{2}{16} =$ $\dfrac{30}{32} =$

$\dfrac{6}{8} =$ $\dfrac{6}{16} =$ $\dfrac{32}{32} =$ $\dfrac{64}{64} =$

$\dfrac{4}{32} =$ $\dfrac{2}{4} =$ $\dfrac{16}{32} =$ $\dfrac{4}{16} =$

(18) Change the mixed number to a fraction:

$\boxed{2\dfrac{1}{2} = \dfrac{5}{2}}$ $3\dfrac{1}{4} = \dfrac{}{4}$ $9\dfrac{1}{2} = \dfrac{}{2}$ $1\dfrac{7}{16} = \dfrac{}{16}$

$4\dfrac{1}{2} = \dfrac{}{2}$ $4\dfrac{1}{16} = \dfrac{}{16}$ $4\dfrac{3}{8} = \dfrac{}{8}$ $8\dfrac{3}{8} = \dfrac{}{8}$

$5\dfrac{1}{4} = \dfrac{}{4}$ $2\dfrac{3}{8} = \dfrac{}{8}$ $10\dfrac{3}{16} = \dfrac{}{16}$ $10\dfrac{1}{8} = \dfrac{}{8}$

(19) Change each fraction to a mixed number:

$\boxed{\dfrac{12}{8} = 1\dfrac{1}{2}}$ $\dfrac{8}{2} =$ $\dfrac{25}{4} =$ $\dfrac{5}{2} =$

$\dfrac{9}{4} =$ $\dfrac{19}{16} =$ $\dfrac{13}{8} =$ $\dfrac{25}{8} =$

$\dfrac{30}{16} =$ $\dfrac{40}{32} =$ $\dfrac{50}{8} =$ $\dfrac{100}{16} =$

(20) Arrange in order from smallest to largest:

$\dfrac{17}{64}$, $\dfrac{2}{4}$, $\dfrac{5}{2}$, $\dfrac{3}{1}$, $\dfrac{8}{8}$, $\dfrac{11}{16}$, $\dfrac{7}{32}$

_____ , _____ , _____ , _____ , _____ , _____ , _____

UNIT 3

ADDITION OF FRACTIONS

INTRODUCTION

Adding fractions involves combining two or more fractions to determine a measure. Before being added, each fraction must have the same denominator and must also be expressed in the same units, for example, inches, millimeters, feet, ounces or pounds. Without a common denominator and unit of measure errors in calculation will occur and unnecessary costs may result.

OBJECTIVES:

After completing this unit the student will be able to:

- Add fractions with the same denominators
- Add fractions and mixed numbers
- Determine the least common denominator (L.C.D.)
- Add fractions with different denominators

ADDITION OF LIKE FRACTIONS:

Fractions having the same denominator are called **LIKE FRACTIONS**.

Such fractions can be easily added together by using the following process:

Step 1   Add the numerators together.

Step 2   Write this sum over the denominator.

Step 3   Reduce the fraction to lowest terms (when possible).

EXAMPLES:

$$\frac{5}{8} + \frac{1}{8} = \underline{\ ?\ }$$

Method A (Horizontal)	Method B (Vertical)
$\xrightarrow{\frac{5}{8} + \frac{1}{8} =}$	$\begin{array}{r}\frac{5}{8}\\ +\frac{1}{8}\end{array}\Bigg\downarrow$

SOLUTION:

Method A:
$$\frac{5}{8} + \frac{1}{8} =$$
$$\frac{5+1}{8} = \frac{6}{8}$$
$$\frac{6 \div 2}{8 \div 2} = \frac{3}{4}$$

Method B:
$$\begin{array}{r}\frac{5}{8}\\ +\frac{1}{8}\end{array}$$
$$\frac{5+1}{8} = \frac{6}{8} = \frac{6 \div 2}{8 \div 2} = \frac{3}{4}$$

NOTE:
   Either Method A or Method B may be used.

✋ The **algebraic form** for ADDITION OF LIKE FRACTIONS is:

$$\frac{a}{c} + \frac{b}{c} = \frac{a+b}{c}$$

💡 **ADDITION OF MIXED NUMBERS**

To add mixed numbers, use this process:

   Step 1   Add the fractions

   Step 2   Add the whole numbers

   Step 3   Reduce the fraction to the lowest terms

   Step 4   Combine the whole numbers with the fractions or mixed number.

EXAMPLE:

$$2\frac{3}{4} + 3\frac{3}{4} = \underline{\ ?\ }$$

$$\begin{array}{r}2\frac{3}{4}\\ +\ 3\frac{3}{4}\\ \hline 5\frac{6}{4} = 5\frac{3}{2}\end{array}\qquad \text{(Change } \tfrac{3}{2} \text{ to a mixed number)}$$

$$= 5 + \frac{3}{2} = 5 + 1\frac{1}{2} = 6\frac{1}{2}$$

ADDITION OF UNLIKE FRACTIONS

Fractions having unlike or different denominators may be added. These fractions are called **UNLIKE FRACTIONS**. Before adding these fractions, it is necessary to change the denominators into a common form such that all denominators are equal.

NOTE:
   Recall the **basic set** of fractions used in the shop. They are:
   $\frac{1}{2}, \frac{1}{4}, \frac{1}{8}, \frac{1}{16}, \frac{1}{32}, \frac{1}{64}$ and multiples of these numbers.

If all fractions being added are from the basic set of **SHOP FRACTIONS**, then they may be changed into **like fractions** having a common denominator equal to the largest denominator in the series. This number is called the **LEAST COMMON DENOMINATOR. (L.C.D.)**

CONSIDER THIS PROBLEM:

$$\frac{3}{4} + \frac{1}{2} + \frac{5}{8} = \underline{\quad ? \quad}$$

All denominators are in the set of shop fractions. "8" is the largest of the three denominators and is, therefore, the L.C.D.

Unlike Fractions		Like Fractions
$\frac{3 \times 2}{4 \times 2}$	=	$\frac{6}{8}$
$\frac{1 \times 4}{2 \times 4}$	=	$\frac{4}{8}$
$\frac{5}{8}$	=	$\frac{5}{8}$

Now we may complete the addition using the rule for **LIKE FRACTIONS**.

Therefore: $\frac{3}{4} + \frac{1}{2} + \frac{5}{8} = \frac{6}{8} + \frac{4}{8} + \frac{5}{8} =$

$$\frac{6 + 4 + 5}{8} = \frac{15}{8} \quad \text{or}$$

$$= 1\frac{7}{8} \quad \text{as a mixed number}$$

If the fractions to be added are not found in the basic set of shop fractions, a different process must be used to determine the **LEASE COMMON DENOMINATOR**. The general process is called **FACTORING**. Factoring requires each denominator to be subdivided into a series of the smallest possible whole numbers which, when multiplied together, will equal the original denominator. Those whole numbers are called **PRIME NUMBERS**.

The set of **PRIME NUMBERS,** numbers with only themselves and unity (1) as factors, include 2, 3, 5, 7, 11, 13, 17, etc.

EXAMPLE 1: $\frac{1}{6} + \frac{1}{9} = \frac{?}{}$

    Step 1 · Write "6" in an equal form using the smallest prime numbers as **factors** of "6." $6 = 2 \times 3$

    Step 2   Express "9" using prime factors. $9 = 3 \times 3$

    Step 3   Combine the factors using **all** factors found in **each number.** However, use each factor the least number of times.

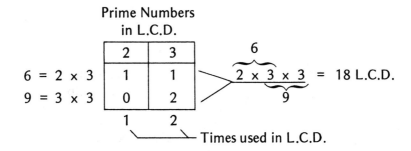

NOTE:
    All factors of **each** number are included, but are used the least number of times.

Therefore: $\frac{1}{6} + \frac{1}{9} = \frac{3}{18} + \frac{2}{18} = \frac{3+2}{18} = \frac{5}{18}$

EXAMPLE 2:

$$\frac{1}{2} + \frac{1}{6} + \frac{1}{12} + \frac{1}{20} = \frac{?}{}$$

Prime Numbers in L.C.D.

	2	3	5
$2 = 1 \times 2$	1	0	0
$6 = 2 \times 3$	1	1	0
$12 = 2 \times 2 \times 3$	2	1	0
$20 = 2 \times 2 \times 5$	2	0	1
	2	1	1

$2 \times 2 \times 3 \times 5 = 60$ L.C.D.

Times used in L.C.D.

$$\frac{1}{2} + \frac{1}{6} + \frac{1}{12} + \frac{1}{20} =$$

$$\frac{30}{60} + \frac{10}{60} + \frac{5}{60} + \frac{3}{60} = \frac{48}{60}$$

Reduce to lowest terms

$$\frac{48 \div 12}{60 \div 12} = \frac{4}{5}$$

THE ALGEBRAIC FORM FOR ADDING ANY TWO FRACTIONS

If only two fractions are being added, you may use an **algebraic process** which is illustrated below:

$$\frac{3''}{4} + \frac{2''}{8} = \underline{\quad?\quad}$$

These numbers follow an algebraic pattern:

$$\frac{a}{b} + \frac{c}{d}$$

 The **rule** is:

$$\frac{a}{b} + \frac{c}{d} = \frac{(a \times d) + (b \times c)}{(b \times d)}$$

or it can be explained as:

$$\frac{a}{b} \searrow \times \frac{}{d} + \frac{}{b} \nearrow \times \frac{c}{d} \text{ over } \overrightarrow{b \times d} = \frac{a}{b} \bowtie \frac{c}{d}$$

Relate the stated problem to the rule:

$$\frac{3}{4} \bowtie \frac{2}{8} \rightarrow \begin{array}{c}\text{Substitute}\\ \text{in Formula}\end{array} \quad \frac{(3 \times 8) + (4 \times 2)}{(4 \times 8)} = \frac{24 + 8}{32} = \frac{32''}{32}$$

which, when simplified to lowest terms, the number $\frac{32}{32} = 1$ or $1''$.

Try this problem:

$$\frac{5''}{8} + \frac{1''}{4} = \underline{\quad?\quad}$$

$$\frac{5}{8} + \frac{1}{4} = \frac{(\quad) + (\quad)}{(\quad)} = \underline{\quad\quad}$$

The correct answer (in lowest terms) is $\frac{7''}{8}$.

CONCEPT APPLICATIONS

Add the fractions and change to mixed numbers in lowest terms.

Set 1

(1) $\dfrac{17}{8} + \dfrac{3}{16} =$ _____

(2) $\dfrac{9}{8} + \dfrac{3}{1} =$ _____

(3) $\dfrac{14}{16} + \dfrac{7}{32} =$ _____

(4) $\dfrac{12}{8} + \dfrac{1}{4} =$ _____

(5) $\dfrac{8}{16} + \dfrac{21}{32} =$ _____

(6) $\dfrac{8}{4} + \dfrac{7}{8} =$ _____

(7) $\dfrac{8}{16} + \dfrac{7}{8} =$ _____

(8) $\dfrac{10}{64} + \dfrac{7}{8} =$ _____

(9) $\dfrac{53}{64} + \dfrac{3}{8} =$ _____

(10) $\dfrac{13}{16} + \dfrac{1}{2} =$ _____

(11) $\dfrac{3}{8} + \dfrac{13}{32} + \dfrac{7}{16} =$ _____

Set 2

(1) $1\dfrac{7}{8}$
$3\dfrac{5}{64}$
$9\dfrac{11}{16}$
_____

(2) $3\dfrac{7}{8}$
$\dfrac{9}{64}$
$1\dfrac{1}{2}$
$8\dfrac{7}{16}$
$\dfrac{11}{32}$
$5\dfrac{3}{4}$
_____

(3) $12\dfrac{3}{4}$
$14\dfrac{5}{8}$
$7\dfrac{5}{16}$
_____

(4) $3\dfrac{1}{2}$
$\dfrac{15}{16}$
$2\dfrac{5}{8}$
_____

(5) $2\dfrac{3}{8}$
$15\dfrac{1}{4}$
$7\dfrac{1}{16}$
_____

(6) $52\dfrac{1}{2}$
$\dfrac{5}{8}$
$\dfrac{13}{16}$
$\dfrac{43}{64}$
_____

Set 3

(1) Find Distance A = _____
     B = _____

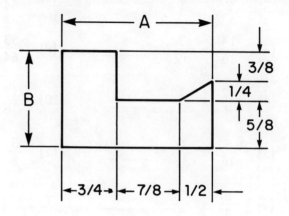

(2) Distance A = _____

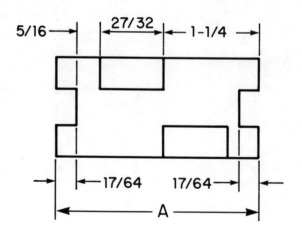

(3)   Distance A = _____

   B = _____

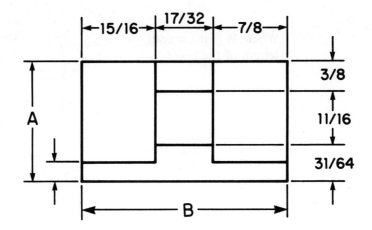

(4)   Measurement A = _____

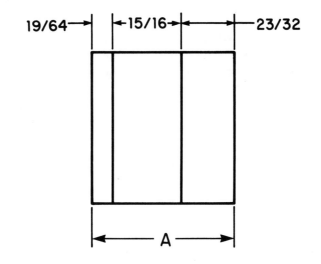

(5)   Distance C = _____

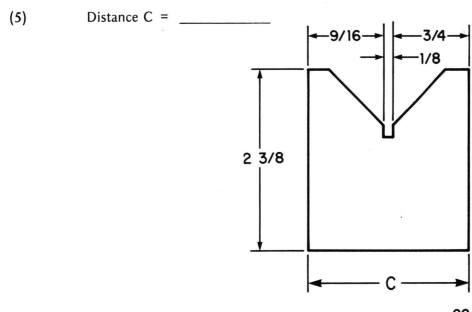

Set 4

(1) What is the overall length of the bolt?

L = _____

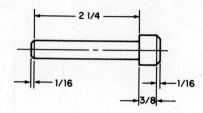

(2) What is the height and width of the block?

H = _____

W = _____

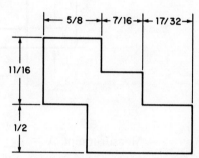

(3) What size piece of sheet metal will be necessary to make the template?

L = _____

W = _____

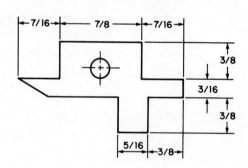

(4) A collar has an inside diameter of $4\frac{3}{8}''$ and a wall thickness of $\frac{3}{32}''$. What is the outside diameter?

O.D. = _____

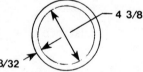

(5) How much wire is required to wind three coil springs if the length needed for each spring is $17\frac{9}{64}''$, $23\frac{5}{8}''$, and $16\frac{9}{32}''$ respectively? Assume a wastage in the cutting of $\frac{1}{8}''$ per spring.

L = _____

(6) Find the Least Common Denominator (L.C.D.) of $\frac{1}{2}$, $\frac{2}{8}$, $\frac{5}{16}$, $\frac{4}{16}$, $\frac{2}{32}$.

L.C.D. = _____

UNIT 4

SUBTRACTION OF FRACTIONS

INTRODUCTION

Subtracting fractions is required to find the difference between two measures or values. Determining the amount of material to be removed from the original stock by a machining operation, positioning a part on a machine during setup or establishing locations for drilling holes, each requires knowledge of the subtraction process.

OBJECTIVES:

After completing this unit the student will be able to:

- Subtract fractions
- Subtract mixed numbers
- Combine addition and subtraction of fractions

SUBTRACTION OF FRACTIONS

The subtraction process is simple when the denominators are the same. That is, when the fractions are **LIKE FRACTIONS**. When **UNLIKE FRACTIONS** are subtracted, the initial process is to express all fractions as **LIKE FRACTIONS** using the **LEAST COMMON DENOMINATOR (L.C.D.)**.

NOTE:

The following terms are used in subtraction:

Example: 7 - 4 = 3   **Terms**

$$\begin{array}{r} 7 \\ -4 \\ \hline 3 \end{array}$$ ← **minuend**
← **subtrahend** (to be subtracted)
← **difference**

The number to be subtracted (the **subtrahend**) is written under the number subtracted from (the **minuend**) and the **difference** is the answer.

For now, the minuend is always the larger number. However, you will later discover, such as in programming applications for numerical control, that this is not always the rule.

The subtraction process is performed with the following:

Step 1   Express the fractions using the L.C.D.

Step 2   Subtract the numerator of the subtrahend from the numerator of the minuend.

Step 3   Write this difference over the denominator.

Step 4   Simplify to lowest terms.

EXAMPLE 1:

$$\frac{13}{16} - \frac{1}{2} = \underline{\quad ? \quad}$$

$$\frac{13}{16} - \frac{1}{2} =$$

$$\frac{13}{16} - \frac{8}{16} = \qquad \text{determine the L.C.D.}$$

$$\frac{13 - 8}{16} = \frac{5}{16}$$

or

$$\frac{13}{16} \qquad \leftarrow \text{minuend}$$

$$-\frac{8}{16} \qquad \leftarrow \text{subtrahend}$$

$$\frac{13 - 8}{16} = \frac{5}{16} \qquad \text{This difference is in } \textbf{Lowest Terms:}$$

32

EXAMPLE 2:

$$\frac{39}{64} - \frac{15}{32} = \underline{\quad ? \quad}$$

$$\frac{39}{64} - \frac{15}{32} =$$

$$\frac{39}{64} - \frac{30}{64} =$$

$$\frac{39 - 30}{64} = \frac{9}{64}$$

or

$$\frac{39}{64} = \frac{39}{64}$$

$$-\frac{15}{32} = -\frac{30}{64}$$

$$\frac{39 - 30}{64} = \frac{9}{64}$$

- - - - - - - - - - - - - - - - - - - -

NOTE:
Remember, when working with shop fractions, the L.C.D. will always be the largest of the given denominators. However, to subtract fractions which are not shop fractions, the L.C.D. must be determined using the **Factoring Process**.

- - - - - - - - - - - - - - - - - - - -

EXAMPLE 3:

$$\frac{41}{56} - \frac{13}{21} = \underline{\quad ? \quad}$$

$56 = 2 \times 2 \times 2 \times 7$
$21 = 3 \times 7$

$2 \times 2 \times 2 \times 3 \times 7 = 168$ L.C.D.

Therefore: $\dfrac{41}{56} - \dfrac{13}{21} = \dfrac{123}{168} - \dfrac{104}{168} = \dfrac{123 - 104}{168} = \dfrac{19}{168}$

ALGEBRAIC FORM FOR SUBTRACTION OF FRACTIONS

As with addition, there is a similar **algebraic process** for **SUBTRACTION OF FRACTIONS** expressed using the following rule:

$$\frac{a}{b} - \frac{c}{d} = \frac{(a \times d) - (b \times c)}{(b \times d)}$$

EXAMPLE 4:

$$\frac{3}{4} - \frac{5}{8} = \underline{\quad ? \quad}$$

$$\frac{3}{4} - \frac{5}{8} = \frac{(3 \times 8) - (4 \times 5)}{(4 \times 8)}$$

$$= \frac{24 - 20}{32} = \frac{4}{32}$$

Always reduce the difference (if possible).

$$\frac{4}{32} = \frac{1}{8}$$

SUBTRACTION OF MIXED NUMBERS

CONSIDER THE FOLLOWING PROBLEM:

$$3\frac{15}{64} - 1\frac{9}{16} = \underline{?}$$

💡 The subtraction process for **mixed numbers** is performed as follows:

Step 1   Change the fractions to like form.

Step 2   If the numerator of the subtrahend (bottom number) is less than the numerator of the minuend, continue subtracting.

Step 3   However, if the numerator of the subtrahend is larger than the numerator of the minuend, change the mixed number to an equal number by adding $\frac{64}{64}$ to $\frac{15}{64}$ and reduce the whole number 3 by 1 unit. This is called **BORROWING**.

Step 4   Reduce when possible.

EXAMPLE 1:

$$3\frac{15}{64} = 3\frac{15}{64} = 2\frac{64}{64} + \frac{15}{64} = 2\frac{79}{64}$$
$$-1\frac{9}{16} = -1\frac{36}{64} = -1\frac{36}{64} = -1\frac{36}{64}$$
$$ 1\frac{43}{64}$$

EXAMPLE 2:

$$4\frac{11}{32} = 3\frac{32}{32} + \frac{11}{32} = 3\frac{43}{32}$$
$$-2\frac{7}{8} = -2\frac{28}{32} = -2\frac{28}{32}$$
$$ 1\frac{15}{32}$$

EXAMPLE 3:

$$13\frac{7}{64} = 12\frac{71}{64}$$
$$-7\frac{5}{16} = -7\frac{20}{64}$$
$$ 5\frac{51}{64}$$

COMBINE ADDITION AND SUBTRACTION

One method of combining addition and subtraction of fractions is to change all mixed numbers into improper fractions, find their L.C.D., and then combine (add or subtract the numerators) as indicated by their signs.

EXAMPLE:

$$2\frac{13}{32} + 4\frac{5}{16} - 3\frac{3}{8} =$$

$$\frac{64+13}{32} + \frac{64+5}{16} - \frac{24+3}{8} = \quad \text{Change to improper fraction}$$

$$\frac{77}{32} + \frac{69}{16} - \frac{27}{8} =$$

$$\frac{77}{32} + \frac{138}{32} - \frac{108}{32} = \quad \text{Find the L.C.D.}$$

$$\frac{77 + 138 - 108}{32} = \quad \text{Combine numerators}$$

$$\frac{215 - 108}{32} =$$

$$\frac{107}{32} = \quad \text{Find lowest terms}$$

$$\frac{96 + 11}{32} = 3\frac{11}{32}$$

CONCEPT APPLICATIONS

Set 1

Subtract: Like Fractions

(1) $\dfrac{5}{16}$ $-\dfrac{1}{16}$ 　(2) $\dfrac{7}{8}$ $-\dfrac{5}{8}$ 　(3) $\dfrac{12}{8}$ $-\dfrac{5}{8}$ 　(4) $\dfrac{13}{16}$ $-\dfrac{5}{16}$ 　(5) $\dfrac{7}{32}$ $-\dfrac{5}{32}$

(6) $2\dfrac{5}{8}$ $-1\dfrac{1}{8}$ 　(7) $3\dfrac{9}{16}$ $-1\dfrac{3}{16}$ 　(8) $8\dfrac{7}{64}$ $-3\dfrac{2}{64}$ 　(9) $4\dfrac{7}{32}$ $-1\dfrac{1}{32}$ 　(10) $7\dfrac{11}{32}$ $-3\dfrac{3}{32}$

(11) $\dfrac{5}{16} - \dfrac{3}{16} =$ _____ 　(12) $\dfrac{11}{64} - \dfrac{9}{64} =$ _____ 　(13) $\dfrac{7}{8} - \dfrac{3}{8} =$ _____

(14) $\dfrac{19}{32} - \dfrac{3}{32} =$ _____ 　(15) $\dfrac{9}{16} - \dfrac{3}{16} =$ _____ 　(16) $\dfrac{7}{8} - \dfrac{5}{8} =$ _____

Subtract: Unlike Fractions

(17) $\dfrac{5}{8}$ $-\dfrac{1}{4}$ 　(18) $\dfrac{1}{2}$ $-\dfrac{1}{8}$ 　(19) $\dfrac{3}{4}$ $-\dfrac{7}{16}$

(20) $\dfrac{3}{4}$ $-\dfrac{5}{32}$ 　(21) $\dfrac{14}{16}$ $-\dfrac{3}{4}$ 　(22) $2\dfrac{5}{16}$ $-1\dfrac{3}{8}$

(23) $3\dfrac{3}{8}$ $-1\dfrac{1}{4}$ 　(24) $2\dfrac{7}{8}$ $-1\dfrac{1}{2}$ 　(25) $5\dfrac{1}{32}$ $-4\dfrac{1}{2}$ 　(26) $7\dfrac{21}{64}$ $-5\dfrac{3}{8}$

Combined (Add and Subtract)

(27) $4 + 3\frac{3}{4} + 7\frac{1}{2} + \frac{1}{8} =$ _____

(28) $\frac{1}{2} + \frac{1}{4} + \frac{1}{8} + \frac{1}{16} + \frac{1}{32} + \frac{1}{64} =$ _____

(29) $2\frac{1}{2} - \frac{3}{4} + \frac{1}{2} - \frac{7}{16} =$ _____

Set 2

(1)   Dimensions A are equal

A = _____

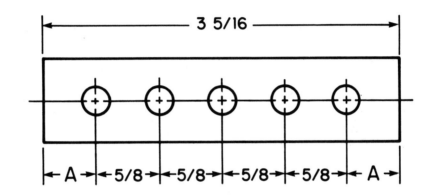

(2)   D = _____

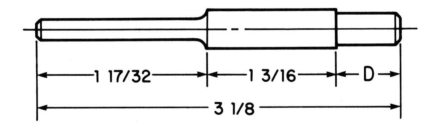

(3)   C = _____

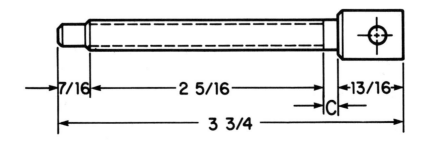

(4)   X = _____

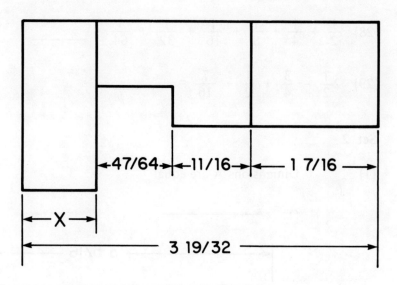

Set 3

(1) Find the difference between 30 and $21\frac{9}{32}$.

(2) A piece of aluminum $4\frac{1}{8}''$ thick was milled to $3\frac{31}{32}''$. What thickness was milled off?

(3) A piece of work on a milling machine is $8\frac{3}{4}''$ thick. It is milled down in 4 cuts of $\frac{1}{4}''$, $\frac{3}{16}''$, $\frac{1}{8}''$, and $\frac{1}{32}''$. What is the thickness of the finished piece?

(4) A section of a part which is $5\frac{1}{8}''$ thick must be machined to $4\frac{19}{32}''$. How thick is the material to be removed?

(5) A nut has an Outside Diameter (O.D.) of $\frac{3}{4}''$ and an Inside Diameter (I.D.) of $\frac{11}{32}''$. What is the thickness of the wall?

(6) If a $\frac{3}{32}''$ cut was taken from $1\frac{1}{4}''$ stock in a lathe, what is the new diameter?

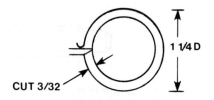

UNIT 5

MULTIPLICATION OF FRACTIONS

INTRODUCTION

Several of the many uses of multiplication include finding areas, volumes and quantities. Multiplication is a process that simplifies and speeds up counting. Using multiplication, the machinist calculates quantities such as material required to do a job, time required for a production run and the number of parts to be produced in a shift.

OBJECTIVES:

After completing this unit the student will be able to:

- Multiply fractions by other fractions
- Multiply fractions by whole numbers
- Multiply fractions by mixed numbers
- Multiply fractions using cancellation

MULTIPLY SIMPLE FRACTIONS BY FRACTIONS

EXAMPLE:

$$\frac{2}{8} \times \frac{5}{4} = \underline{\quad ? \quad}$$

$$\boxed{\frac{2}{8}} \times \boxed{\frac{5}{4}} = \frac{2 \times 5}{8 \times 4} = \boxed{\frac{10}{32}} \longleftarrow \text{Product}$$

Factors

$$\frac{10}{32} = \frac{5}{16} \longleftarrow \text{Remember to Reduce}$$

NOTE:
The numbers being multiplied are called **FACTORS**. The answer or result is a **PRODUCT**.

39

The algebraic form for **MULTIPLICATION OF FRACTIONS BY FRACTIONS** is:

$$\frac{a}{b} \times \frac{c}{d} = \frac{a \times c}{b \times d} = \frac{ac}{bd}$$

MULTIPLY A FRACTION BY A WHOLE NUMBER

CONSIDER THE FOLLOWING

$$\frac{7}{16} \times 3 = \underline{?}$$

NOTE:

(1) Numbers may be multiplied in any order. $\frac{7}{16} \times 3 = 3 \times \frac{7}{16}$

(2) Whole numbers can be written as fractions using "1" as the denominator. $3 = \frac{3}{1}$

EXAMPLE:

$$\frac{7}{16} \times 3 = \frac{7}{16} \times \frac{3}{1} =$$

$$\frac{7 \times 3}{16 \times 1} = \frac{21}{16} \quad \text{Reduce}$$

$$= \frac{16 + 5}{16} = 1\frac{5}{16}$$

MULTIPLY A FRACTION BY A MIXED NUMBER

The process is performed as follows:

Step 1   Change each mixed number factor to an improper fraction

Step 2   Multiply like any other pair of fractions

Step 3   Reduce to lowest terms

EXAMPLE 1:

$$2\frac{5}{8} \times 1\frac{1}{2} = \underline{?}$$

$$2\frac{5}{8} \times 1\frac{1}{2} = \frac{21}{8} \times \frac{3}{2} = \frac{21 \times 3}{8 \times 2} = \frac{63}{16} = \frac{48 + 15}{16} = 3\frac{15}{16}$$

EXAMPLE 2:

$$3\frac{1}{16} \times 4\frac{1}{4} = \underline{?}$$

$$3\frac{1}{16} \times 4\frac{1}{4} = \frac{49}{16} \times \frac{17}{4} = \frac{49 \times 17}{16 \times 4} = \frac{833}{64} = \frac{832 + 1}{64} = 13\frac{1}{64}$$

CANCELLATION LAW IN MULTIPLICATION

A process called **CANCELLATION** is used to simplify a fraction to lower terms or to simplify the fractions involved in a multiplication operation.

OBSERVE:

$$\frac{n}{n} = 1, \quad \frac{2}{2} = 1, \quad \frac{8}{8} = 1, \quad \frac{10}{5} = \frac{2}{1} = 2$$

Compare the processes used in the following example:

(a) Multiply first, then cancel:

$$\frac{16}{2} \times \frac{8}{4} \times \frac{2}{16} \times \frac{16}{8} = \frac{16 \times 8 \times 2 \times 16}{2 \times 4 \times 16 \times 8} = \frac{4096}{1024} = \frac{4}{1} = 4$$

(b) or, Cancel first, then multiply:

$$\frac{\cancel{16}}{\cancel{2}} \times \frac{\cancel{8}}{4} \times \frac{\cancel{2}}{\cancel{16}} \times \frac{16}{\cancel{8}} = \frac{16}{4} = \frac{4}{1} = 4$$

OBSERVE:

Any numerator factor may cancel an equal denominator factor. This **CANCELLATION** process greatly simplified the multiplication of part (b). The enclosed numbers in part (b) cancelled each other in both the numerator and denominator:

$$\frac{16}{16}, \frac{8}{8}, \frac{2}{2} = \frac{1}{1} = 1 \quad \text{leaving the product} \quad \frac{16}{4} = 4$$

✋ The **CANCELLATION LAW** states that any factor in a numerator(s) may cancel an equal factor in a denominator(s).

The cancellation process is used to cancel factors other than those that equal "1".

CONSIDER THE FOLLOWING:

EXAMPLE 1:

$\frac{12}{8}$ which could be written as:

$$\frac{12}{8} = \frac{3 \times 4}{2 \times 4} = \frac{3}{2} \times \frac{\cancel{4}}{\cancel{4}} = \frac{3}{2} \times 1 = \frac{3}{2} = 1\frac{1}{2}$$

or simplified:

$$\frac{12}{8} \quad \text{as} \quad \frac{\cancel{12}^3}{\cancel{8}_2} = \frac{3}{2} = 1\frac{1}{2}$$

OBSERVE:

A factor of "4" has been removed from both the numerator, "12", and the denominator, "8", by cancellation.

EXAMPLE 2:

$$\frac{30}{8} \times \frac{2}{10} \times \frac{4}{16} = \underline{\ ?\ }$$

$$\frac{\overset{3}{\cancel{30}}}{8} \times \frac{2}{\underset{1}{\cancel{10}}} \times \frac{4}{16} \quad \text{Cancel } \frac{10}{10} \quad = \frac{3}{8} \times \frac{2}{1} \times \frac{4}{16} \quad \left(\text{Note: } \frac{30}{10} = \frac{3}{1}\right)$$

$$\frac{3}{\underset{2}{\cancel{8}}} \times \frac{2}{1} \times \frac{\overset{1}{\cancel{4}}}{16} \quad \text{Cancel } \frac{4}{4} \quad = \frac{3}{2} \times \frac{2}{1} \times \frac{1}{16} \quad \left(\text{Note: } \frac{4}{8} = \frac{1}{2}\right)$$

$$\frac{3}{\cancel{2}} \times \frac{\cancel{2}}{1} \times \frac{1}{16} \quad \text{Cancel } \frac{2}{2} \quad = \frac{3}{1} \times \frac{1}{1} \times \frac{1}{16} \quad \left(\text{Note: } \frac{2}{2} = \frac{1}{1}\right)$$

$$= \frac{3}{16}$$

With practice you will learn to reduce the number of steps in the above process as follows:

EXAMPLE:

$$\frac{7}{32} \times \frac{40}{49} = \underline{\ ?\ }$$

$$\frac{\overset{1}{\cancel{7}}}{\underset{4}{\cancel{32}}} \times \frac{\overset{5}{\cancel{40}}}{\underset{7}{\cancel{49}}} = \frac{5}{28} \qquad \text{Cancel "7" in ("7" and "49"),} \\ \text{and "8" in ("32" and "40").}$$

SUMMARY OF MULTIPLICATION OF FRACTIONS

Step 1 Change all factors to common fractions

Step 2 Use cancellation process to simplify whenever possible

Step 3 Multiply all numerators

Step 4 Multiply all denominators

Step 5 Express the resulting common fraction in lowest terms

― ― ― ― ― ― ― ― ― ― ― ― ― ― ― ― ― ―

NOTE:
Step 5 is only necessary if the cancellation process was **not** completed in Step 2.

― ― ― ― ― ― ― ― ― ― ― ― ― ― ― ― ― ―

CONCEPT APPLICATIONS:

Set 1: Multiply

(1) $\dfrac{1}{4} \times 7 =$ _____ (2) $\dfrac{1}{2} \times 8 =$ _____ (3) $\dfrac{1}{8} \times \dfrac{16}{1} =$ _____

(4) $\dfrac{1}{2} \times \dfrac{8}{1} =$ _____ (5) $\dfrac{3}{4} \times \dfrac{5}{8} =$ _____ (6) $\dfrac{1}{2} \times \dfrac{5}{8} =$ _____

(7) $\dfrac{3}{8} \times \dfrac{1}{2} =$ _____ (8) $\dfrac{5}{8} \times \dfrac{2}{3} =$ _____ (9) $\dfrac{4}{5} \times \dfrac{2}{3} =$ _____

(10) $\dfrac{3}{4} \times \dfrac{4}{3} =$ _____ (11) $\dfrac{5}{8} \times \dfrac{8}{16} =$ _____ (12) $\dfrac{1}{64} \times \dfrac{8}{16} =$ _____

(13) $2\dfrac{1}{2} \times \dfrac{1}{4} =$ _____ (14) $1\dfrac{1}{8} \times \dfrac{3}{4} =$ _____ (15) $2\dfrac{1}{2} \times 3\dfrac{1}{4} =$ _____

(16) $3\dfrac{1}{4} \times 3\dfrac{1}{4} =$ _____ (17) $3\dfrac{1}{2} \times 4\dfrac{1}{4} =$ _____ (18) $4\dfrac{1}{4} \times \dfrac{1}{16} =$ _____

(19) $1\dfrac{1}{4} \times 8\dfrac{1}{8} =$ _____ (20) $19 \times 1\dfrac{1}{2} =$ _____ (21) $8\dfrac{3}{4} \times 25 =$ _____

(22) $\dfrac{3}{8} \times \dfrac{1}{2} \times \dfrac{4}{5} =$ _____ (23) $\dfrac{2}{5} \times \dfrac{4}{9} \times \dfrac{15}{16} =$ _____ (24) $\dfrac{3}{8} \times \dfrac{5}{6} \times 12 =$ _____

(25) $1\dfrac{1}{2} \times \dfrac{4}{5} \times 2\dfrac{1}{4} =$ _____ (26) $3\dfrac{1}{4} \times 3\dfrac{1}{2} \times 4 =$ _____ (27) $1\dfrac{1}{3} \times 3\dfrac{1}{4} \times 4\dfrac{1}{8} =$ _____

* Challenge Problems:

(1) $\dfrac{1}{2} \times \dfrac{1}{4} \times \dfrac{1}{8} \times \dfrac{1}{16} \times \dfrac{1}{32} \times \dfrac{1}{64} =$ _____

(2) $\dfrac{64}{2} \times \dfrac{16}{4} \times \dfrac{4}{8} \times \dfrac{32}{16} \times \dfrac{2}{32} \times \dfrac{8}{64} =$ _____

Set 2

(1) $8 \times \dfrac{3}{8}$

(2) $\dfrac{3}{4} \times \dfrac{7}{8}$

(3) $7\dfrac{1}{8} \times 2\dfrac{1}{2}$

(4) What is $\dfrac{1}{4}$ of 28? (Yes, "of" means "times" or $\dfrac{1}{4} \times 28$.)

(5) $10 \times \dfrac{3}{4} \times \dfrac{3}{8}$

(6) If the taper of a round shaft is $\dfrac{3}{4}$ inch per foot, how much does the diameter decrease in 9 inches?

(7) A production of 3000 parts per week are to be delivered to the vendor. Due to production problems the full production could not be met. The manufacturer shipped $\dfrac{3}{4}$ of the parts for that week. How many parts were on back order?

(8) If the blanking pressure for making a blank out of mild steel is equal to $25\dfrac{1}{2}$ tons per square inch of cross section, what would be the tonnage required to blank out a part that is $\dfrac{1}{8}''$ thick and a length of cut of $13\dfrac{1}{2}$ inches. *Note:* The cross section in square inches is equal to the thickness of the material times the length of the cut.

(9) What is the area of a triangle that is $3\dfrac{1}{4}$ at the base with a height (H) of $4\dfrac{1}{2}$. ($A = \dfrac{1}{2} B \times H$)

(10) Six steel blocks $5\dfrac{1}{4}$ inches by $6\dfrac{3}{4}$ inches are to be placed in an oven that is $10\dfrac{1}{2}$ inches wide. How long should the area be?

(11) The circumference of a circle is about $3\dfrac{1}{7}$ times the diameter of the circle. If the diameter is equal to 21 inches, what is the circumference?

(12) If two pieces of metal are to be bolted together with a bolt $3\frac{1}{4}$ inches long and the bolt is to be threaded into the one part a depth equal to $\frac{1}{8}$ the length of the bolt, what depth of tapped hole will be needed?

(13) A bar of metal weighs $2\frac{3}{8}$ pounds for each foot of length. How much will a piece weigh, if it is 7 feet 8 inches long?

(14) What length of stock would be required to make 25 taps each $3\frac{5}{8}''$ long if each one is allowed $\frac{1}{16}''$ for waste?

(15) How many pounds of metal are in a production run of 350 pieces, if each piece weighs 1 pound 14 ounces?

UNIT 6

DIVISION OF FRACTIONS

INTRODUCTION

Division of fractions is the process of determining how many parts there are in some larger quantity. For instance, how many parts of fractional length are in a bar of steel stock, or how many molded parts can be produced from a barrel of epoxy?

OBJECTIVES:

After completing this unit the student will be able to:

- Divide: fractions by fractions
  fractions by whole numbers
  fractions by mixed numbers

- Divide: whole numbers by fractions
  whole numbers by mixed numbers

- Divide: mixed numbers by whole numbers
  mixed numbers by fractions
  mixed numbers by mixed numbers

- Combine multiplication and division of fractions in the same problem

- Convert multiplication to division and division to multiplication

- Simplify complex fractions

DIVISION is an operation which allows us to determine how many times one quantity is found in another. This idea is expressed as follows:

CONSIDER:
How many times is 4 contained in 32?

$$\text{divisor} \longrightarrow 4 \overline{\smash{)}32} \longleftarrow \text{quotient} \atop \text{dividend}$$
$$\underline{32}$$
$$0 \longleftarrow \text{remainder}$$

NOTE:
The **product** of "8" and "4" plus the remainder (in this case, zero) is equal to the **dividend**. Therefore, multiplication and division are **OPPOSITE** or **INVERSE** operations.

$$4 \times 8 = 32 \quad \text{while} \quad 32 \div 4 = 8$$
$$\text{and} \quad 32 \div 8 = 4$$

CONSIDER:

$$10 \div 2 = 5 \quad \text{is the same as} \quad 10 \times \frac{1}{2} = 5$$

and $\quad 12 \div 4 = 3 \quad \longrightarrow \quad 12 \times \frac{1}{4} = 3$

and $\quad 24 \div 3 = 8 \quad \longrightarrow \quad 24 \times \frac{1}{3} = 8$

OBSERVE:
Instead of dividing by 2, 4, and 3, we get the same result when we multiply by $\frac{1}{2}$, $\frac{1}{4}$ and $\frac{1}{3}$.

Numbers of this type: 2 and $\frac{1}{2}$,
3 and $\frac{1}{3}$,
4 and $\frac{1}{4}$, etc. are called **RECIPROCALS** or **INVERSES** of each other.

Other Examples:

$\frac{4}{5}$ and $\frac{5}{4}$, $\quad 2\frac{1}{2} = \frac{5}{2}$ and $\frac{2}{5}$ $\quad \frac{7}{8}$ and $\frac{8}{7}$,

RULE:
We may solve any division problem if we change it into a multiplication problem by:

Step 1 Changing the " $\div$ " division sign to a " $\times$ " multiplication sign

Step 2 Changing the divisor into its reciprocal

Step 3 Following the rule for multiplication of fractions from page 40.

47

NOTE:
> Mixed numbers first must be changed to improper fractions before inverting them.

EXAMPLE 1:

$$\frac{3}{8} \div \frac{1}{2} = \underline{\quad ? \quad}$$

$$\frac{3}{8} \div \frac{1}{2} =$$

$$\frac{3}{8} \left(\div \frac{1}{2} \right) =$$

$$\frac{3}{8} \times \frac{2}{1} = \frac{6}{8} = \frac{3}{4}$$

Remember, when you see this (÷), do this "invert the **second** fraction and multiply."

EXAMPLE 2:

$$5\frac{7}{16} \div \frac{29}{32} =$$

$$\frac{80 + 7}{16} \div \frac{29}{32} =$$

$$\frac{\overset{3}{\cancel{87}}}{\underset{1}{\cancel{16}}} \times \frac{\overset{2}{\cancel{32}}}{\underset{1}{\cancel{29}}} = \frac{6}{1} = 6$$

EXAMPLE 3:
> How many "$\frac{1}{2}$ inch units" exist in 1 foot or 12 inches? This is determined as follows:

$$12 \div \frac{1}{2} =$$

$$12 \times \frac{2}{1} = \frac{24}{1} = 24 \quad \text{There are 24 units, each } \tfrac{1}{2}'' \text{ long in 1 foot.}$$

The **algebraic process** for **DIVISION OF FRACTIONS** is

RULE:

$$\frac{a}{b} \div \frac{c}{d} = \frac{ad}{bc}$$

OBSERVE:

$$\frac{a}{b} \div \frac{c}{d} = \frac{a}{b} \times \frac{d}{c} = \frac{a \times d}{b \times c} = \frac{ad}{bc}$$

As you become familiar with the process, you can apply a short cut procedure to find the answer in one step:

$$\frac{a}{b} \div \frac{c}{d} = \frac{ad}{bc}$$ (Sometimes called a form of **cross multiplication**.)

EXAMPLE 1:

$$\frac{3}{5} \div \frac{2}{3} = \frac{9}{10}$$ in 1 easy step.

EXAMPLE 2:

$$\frac{3}{4} \div \frac{5}{2} = \frac{6}{20}$$ which can be reduced to $\frac{3}{10}$.

✋ A **COMPLEX FRACTION** is one where either the numerator, the denominator, or both numerator and denominator are given as a fraction or mixed number. All such fractions require the use of the division rules to simplify.

CONSIDER:

$$\frac{8}{\frac{3}{4}} \,,\; \frac{\frac{1}{2}}{\frac{1}{8}} \,,\; \frac{2\frac{1}{2}}{\frac{1}{16}}$$

These complex fractions can be expressed in a simple form as:

$$(8 \div \frac{3}{4}),\; (\frac{1}{2} \div \frac{1}{8}),\; (2\frac{1}{2} \div \frac{1}{16}),$$ and then divided or simplified.

EXAMPLE 1:

$$\frac{8}{\frac{3}{4}} = 8 \div \frac{3}{4} = 8 \times \frac{4}{3} = \frac{32}{3} = 10\frac{2}{3}$$

EXAMPLE 2:

$$\frac{\frac{1}{2}}{\frac{1}{8}} = \frac{1}{2} \div \frac{1}{8} = \frac{1}{2} \times \frac{8}{1} = \frac{4}{1} = 4$$

EXAMPLE 3:

$$\frac{2\frac{1}{2}}{\frac{1}{16}} = 2\frac{1}{2} \div \frac{1}{16} = \frac{5}{2} \times \frac{16}{1} = \frac{40}{1} = 40$$

CONCEPT APPLICATIONS

Set 1: Divide the following fractions and reduce if possible:

(1) $\dfrac{1}{2} \div \dfrac{3}{4} =$

(2) $\dfrac{3}{4} \div \dfrac{1}{2} =$

(3) $\dfrac{3}{10} \div \dfrac{3}{5} =$

(4) $\dfrac{3}{10} \div \dfrac{33}{100} =$

(5) $\dfrac{48}{64} \div \dfrac{16}{32} =$

(6) $\dfrac{3}{1000} \div \dfrac{1}{10} =$

(7) $\dfrac{3}{15} \div \dfrac{21}{65} =$

(8) $\dfrac{3}{10000} \div \dfrac{3}{2} =$

(9) $\dfrac{3}{16} \div \dfrac{6}{8} =$

(10) $\dfrac{1}{16} \div \dfrac{1}{64} =$

Set 2

(1) Divide $\dfrac{3}{4}$ by 3

(2) Divide 8 by $\dfrac{1}{2}$

(3) Divide 9 by $2\dfrac{1}{2}$

(4) Divide $15\dfrac{1}{4}$ by $2\dfrac{1}{2}$

(5) $27\dfrac{3}{4} \div 9\dfrac{1}{4} =$

(6) $\dfrac{3}{4} \div \dfrac{1}{2} =$

(7) $4\dfrac{11}{16} \div 4 =$

(8) Allowing $\dfrac{1}{8}''$ for cutting off each nut, how many nuts $\dfrac{3}{8}''$ thick can be cut from a piece of cold rolled steel 12 feet long?

(9) Divide a bar 7 feet 6 inches into 6 equal parts.

Set 3

(1) Divide a bar 13 feet long into equal parts $3\frac{1}{4}$ inches long. How many parts are produced?

(2) There are 75 pounds of bolts in a bin. If each bolt weighs $\frac{3}{8}$ pounds, how many bolts are there in the bin?

(3) How many $\frac{3}{16}''$ thick washers can be cut from a piece of bar stock $\frac{3}{16}''$ diameter and $5\frac{1}{16}''$ long? (*Note:* Do not consider the material lost in the sawing operation.)

(4) If a bar weighs $2\frac{1}{4}$ lb. per foot, how long must it be to weigh 36 lbs?

(5) Find the number of threads of $\frac{3}{64}''$ pitch of a bolt threaded $3\frac{3}{4}''$. The pitch of the thread is the distance from one thread to the next. The formula for the number of threads (N): $N = \frac{1}{\text{pitch}}$

(6) If the lead of the thread (the distance the nut advances in one turn) on a bolt is $\frac{3}{32}''$, how many turns will a nut make in order to travel $\frac{7}{8}''$?

(7) How many metal sheets, each $\frac{1}{32}''$ thick are there in a stack $25\frac{1}{2}''$ high?

(8) Allowing $\frac{1}{8}''$ for each cut, how many $3\frac{1}{2}''$ sections can be cut from a bar $9\frac{2}{3}$ ft. long?

(9) A screw has a pitch of $\frac{3}{16}''$. Find the Number of threads per inch. $N = \frac{1}{\text{pitch}}$

(10) If the thickness of a fastener head is diameter ÷ 4, and the length of the shaft is diameter × $1\frac{1}{8}$, find the head thickness and shaft length when the diameter is $\frac{1}{4}''$.

(11) Seven holes are to be equally spaced, in line, on a piece of stock $1'\text{-}10\frac{1}{4}''$ long. The center of the outermost holes on each end will be $1\frac{1}{4}''$ from the end of the stock. What will be the distance between centers of each hole? *Hint:* make a sketch first.

51

UNIT 7

DECIMALS

INTRODUCTION

In Unit 2, a special set of fractions were identified which have since been referred to as "**shop fractions.**" They are: $\frac{1}{2}, \frac{1}{4}, \frac{1}{8}, \frac{1}{16}, \frac{1}{32}, \frac{1}{64}$. These fractions were found useful in increasing the accuracy of measurement. The concepts of equivalent (=) fractions and the operations of addition, subtraction, multiplication and division were developed. Now, to further increase the ability to measure with greater precision, a similar set of numbers called decimals will be studied. Again, look for equivalent names for each number, find a rule to perform each of the basic operations, (+, −, x, ÷), and become familar with the applications of decimals to actual problem solving in the shop.

OBJECTIVES:

After completing this unit the student will be able to:

- Express decimals in word form and in fractional form
- Locate the decimal point in a mixed decimal
- Locate a decimal on a number scale

The **DECIMAL SYSTEM*** consists of a set of numbers taken from our general set of fractions. Specifically, they are those fractions that contain a denominator of 10, 100, 1,000 and so on:

such as $\quad \frac{1}{10}, \frac{1}{100}, \frac{1}{1000}, \frac{1}{10000}$

written as: $\quad$ .1, .01, .001, .0001

💡 These denominators are multiples (powers) of "10." The decimal place of each fraction is decreased 10 times to get the next decimal place. This is easily shown using the equivalent values with dollars and cents.

*The term "decimal" is normally used to mean .1, .01, etc. and the term "fraction" to mean $\frac{1}{2}$, $\frac{1}{4}$, etc. More accurately, the terms **decimal fractions** and **common fractions** should be used. A fraction is defined as an equal part or number of parts of a whole unit and both forms fit this definition. However, the **DECIMAL SYSTEM** is based on the unique set of common fractions which have denominators of 10, 100, 1000, etc. and this set of fractions we will call **decimals**. Common fractions in the form of $\frac{n}{d}$ will be referred to simply as **fractions**.

EXAMPLE:

$\frac{1}{10}$ of a dollar = 10 cents = $.10

$\frac{1}{100}$ of a dollar = 1 cent = $.01

Therefore, .1 has 10 times the value of the next number, .01. **Agreements** have been made which allow us to abbreviate or simplify the form of these fractions. To understand the concept of writing these numbers we must first examine the following chart which shows how **PLACE VALUE** is assigned to each digit. Always begin with the identification of the "Ones" or "Units" Place. The "**Point**" is located to the right of the "**Unit**."

PLACE VALUE CHART

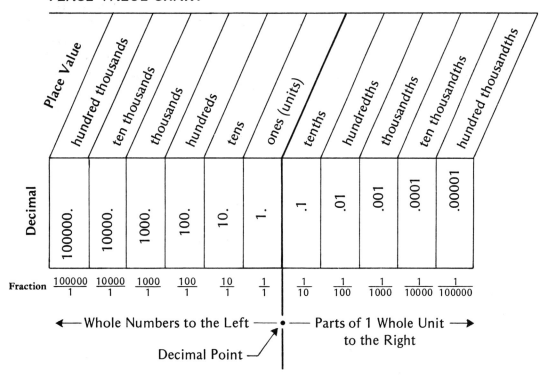

EXAMPLES OF EQUIVALENT FORMS:

Decimal Form: 1 2 3 4 • 5 6 7

Fraction Form: 1, 2 3 4 $\frac{567}{1000}$

Word Form: One thousand, two hundred thirty-four and five hundred sixty-seven thousandths.

Write word names for the following:

a. .45   b. 9.02   c. 372.253   d. .9   e. .0009

SOLUTIONS:

a. Forty-five hundredths
b. Nine and two hundredths
c. Three hundred seventy-two and two hundred fifth-three thousandths
d. Nine tenths*
e. Nine ten-thousandths

Write fraction or mixed number names for these decimals:

a. .4   b. 25.001   c. 300.32

SOLUTIONS:

a. $.4 = \frac{4}{10} = \frac{2}{5}$

b. $25.001 = 25\frac{1}{1000}$

c. $300.32 = 300\frac{32}{100} = 300\frac{8}{25}$

💡 A machinist must constantly compare measures. **Comparison** of decimals is very easy when they have the **same** number of decimal places. For example, the numbers .25, .44, and .99. (See scale below). It should be obvious that 25 hundredths is less than 44 hundredths. Also, 44 hundredths is less than 99 hundredths. This relationship may also be expressed on a number scale which identifies 100 units or subunits.

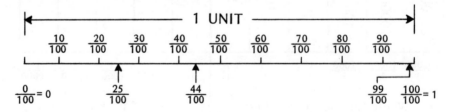

> *The term "a tenth" is commonly used in machine shop terminology relating to precision measures. A tenth is $\frac{1}{10000}$ or .0001. A measure of "two tenths" may be used to describe a dimension to an accuracy of .0002. We do not apply this use of the term in Practical Mathematics, but in precision machine shop applications you will learn how and when to use the term meaningfully.

To compare decimals not having an equal number of places:

.7 .65 .615

We annex zeros to these numbers so that they will have the **same** number of places.

NOTE:
Annexation of zeros does not change the actual value of the decimal number.

EXAMPLE: .7 = .70 = .700
.65 = .650
.615   then,

Compare these numbers in terms of a **common unit**, which would be thousandths. When arranged in order of smallest to largest —

.615 .650 .700

These decimals would exist in this order on a number scale, (small-to-large) with the larger number always found to the **right** side of the small number.

55

CONCEPT APPLICATIONS

Set 1 Write each decimal in word form: (Refer to the Place Value Chart.)

(1) .011 _____

(2) .12 _____

(3) .246 _____

(4) .3 _____

(5) .516 _____

(6) .2185 _____

(7) .001 _____

(8) .0012 _____

(9) 35.07 _____

(10) 2.222 _____

(11) 30.003 _____

(12) 350.5 _____

(13) .00022 _____

(14) .00101 _____

(15) 7.654321 _____

Set 2 Write the following in decimal form:

(1) One tenth _____

(2) Two hundred and fifty thousandths _____

(3) Thirty-two ten-thousandths _____

(4) Three and twenty five thousandths _____

(5) Twenty-five and six hundred six thousandths _____

Set 3 Write the following in fraction form:

(1) .5 _____   (2) .06 _____   (3) .007 _____

(4) .0008 _____   (5) .25 _____   (6) .250 _____

(7) .2500 _____   (8) 45.4 _____   (9) 7.69 _____

(10) 34.125 _____

Set 4 Arrange in order — small to large:

(1) .6   .065   .63   _____ _____ _____

(2) .026   .25   .2   _____ _____ _____

(3) .6   .666   .66   _____ _____ _____

(4) .027   2.7   .2702   _____ _____ _____

(5) .097   .0938   .3   _____ _____ _____

UNIT 8

ROUNDING OFF DECIMALS

INTRODUCTION

Precision machinery and the accuracy of measuring equipment enable the machinist to produce and measure parts with a precise mirror finish at a certain fine surface texture. Though such precision machining is feasible, it is costly and often is not required. Pocket calculators compute dimensions with much greater accuracy than is normally needed to produce an acceptable part. Round off decimals so that parts meet design specifications without greatly exceeding them to increase productivity and reduce costs.

OBJECTIVES:

After completing this unit the student will be able to:

- Round off decimals to a required decimal place

When the standard operations of addition, subtraction, etc. are performed on a set of decimals, the results are frequently numbers which relate to measurements which are awkward to work with. For instance, if instructions are to cut a bolt exactly 3.8547621 inches long, the chances are good that the instructions on this job order are a mistake. There are no instruments which a machinist would use to determine the accuracy of such a measurement. The standard shop practice is to work with decimal measures to only three places. Therefore, the number, 3.8547621 would be simplified and expressed using only 3 decimal places. Although this new number is no longer exactly equal to the original number, the value is close enough for general shop work. The process of simplifying a decimal is referred to as **"ROUNDING OFF."**

The general rule for **ROUNDING OFF** a decimal is:

    a. Determine the degree of accuracy required in terms of the specific place (such as thousandths).

    b. If the digit following the thousandths place is less than 5, drop all digits to the right of the required place (thousandths).

    c. If the digit following the thousandths place is greater than 5, add one to the thousandths digit and drop all digits to the right of the required place (thousandths).

    d. If the digit following the thousandths place is exactly 5, round off to the **nearest even digit** in the thousandths place.

EXAMPLE 1: Round off .743429 to three places or thousandths,

— The digit following the third place is 4.

— Since 4 is less than 5, drop all digits after the third place.

Therefore: The result, .743 is said to be rounded off and is accurate to the nearest thousandth.

— — — — — — — — — — — — — — — — — — —

NOTE:

The general rule applies for ROUNDING OFF to any specified number of places. The third, or thousandths, place was used to simplify the illustration.

— — — — — — — — — — — — — — — — — — —

An easy to understand presentation of rounding off is a comparison with an automobile odometer reading 62508.8 miles. Assume this is a decimal with the decimal point at the far left. We want to round off to four places or ten thousandths:

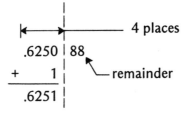

Since the remainder, "88," is greater than "5," add "1" to the fourth place to get ".6251."

If the remainder had been "48," we would round down to .6250. Even though the odometer had passed that number, it has not yet reached the halfway point to the next higher place.

59

EXAMPLE 2: Round off .56835 to **two** places.

- The digit following the second place is 8.

- Since 8 is larger than 5, add one to the required second place digit, and then drop the remaining digits located to the right of the second place digit. Therefore, the result, .57 is rounded off and accurate to the nearest **hundredth**.

EXAMPLE 3: Round off .78345 to:

1 place (tenth)	.8
2 places (hundredth)	.78
3 places (thousandth)	.783
4 places (ten thousandth)	.7834 (refer to part d of the rule)

EXAMPLE 4: Round off .8735 to the nearest thousandth .874

Round off .8745 to the nearest thousandth _____

CONCEPT APPLICATIONS

Round off the following to three places (thousandths). ($\approx$) means rounded off or approximately equal.

(1) .7874 $\approx$ _____

(2) .07874 $\approx$ _____

(3) .23622 $\approx$ _____

(4) 2.1653565 $\approx$ _____

(5) 3.58267 $\approx$ _____

(6) .42103 $\approx$ _____

(7) .4215 $\approx$ _____

(8) .42193 $\approx$ _____

(9) .17185 $\approx$ _____

(10) .00357 $\approx$ _____

UNIT 9

CHANGING FRACTIONS TO DECIMALS

INTRODUCTION

When dimensions on shop prints are given in common shop fractions, the greatest accuracy is $\frac{1}{64}$ of an inch. This is not precise enough (.016″) for many machining operations. Therefore, when fractions are given on a blueprint, they must often be converted to decimals before machining operations begin so that greater accuracy can be obtained.

OBJECTIVE:

After completing this unit the student will be able to:

- Change a common fraction to its decimal equivalent.

CHANGING A FRACTION INTO A DECIMAL FORM:

Step 1 Divide the numerator of the fraction by the denominator.

Step 2 Round off the answer to the required number of places.

EXAMPLE 1: Change $\frac{3}{8}$ to a decimal.

$$
\begin{array}{r}
.375 \\
8 \overline{\smash{)}3.000} \\
\underline{24} \\
60 \\
\underline{56} \\
40 \\
\underline{40}
\end{array}
$$

or $\frac{3}{8} = .375$

OBSERVE: The answer came out **even** to thousandths place (zero remainder). Compare Example 1 with the following:

61

EXAMPLE 2: Change $\frac{1}{3}$ to a decimal.

```
        .3333 ....
    ┌─────────
  3 ) 1.0000
       9
       ──
       10
        9
        ──
        10
         9
         ──
         10
          9
```

OBSERVE: The answer does not come out even (a remainder "1" continues).

─ ─ ─ ─ ─ ─ ─ ─ ─ ─ ─ ─ ─ ─ ─ ─ ─ ─

NOTE:
When this type pattern is observed, we simply stop the repeated division steps and round off the answer to the required number of places. Thus,

$\frac{1}{3}$ = .333 (in thousandths) = .33 (in hundredths) = .3 (in tenths)

─ ─ ─ ─ ─ ─ ─ ─ ─ ─ ─ ─ ─ ─ ─ ─ ─ ─

All of the basic **SHOP FRACTIONS** when changed to a decimal form do not have a repeating pattern. That is, the remainder becomes zero.

Change the following fractions to decimal form using as many steps as necessary to have the answer come out even (zero remainder). Do your answers agree with those given?

a. $\frac{1}{2}$ = .5   d. $\frac{1}{16}$ = .0625

b. $\frac{1}{4}$ = .35   e. $\frac{1}{32}$ = .03725

c. $\frac{1}{8}$ = .125   f. $\frac{1}{64}$ = .015625

Which of these answers are incorrect? _____ and _____

CONCEPT APPLICATIONS

Reduce the following fractions to lowest terms and express each as a decimal.

(1) $\dfrac{28}{32}$ = $\dfrac{7}{8}$ = .875

(2) $\dfrac{14}{32}$ = —— = _____

(3) $\dfrac{6}{16}$ = —— = _____

(4) $\dfrac{15}{40}$ = —— = _____

(5) $\dfrac{18}{64}$ = —— = _____

NOTE:
When changing a fraction to decimal form, the division is simplified if the fraction is reduced to lowest terms and then perform the division operation:

$\dfrac{48}{64}$ → $\begin{array}{r} .75 \\ 64\overline{)48.00} \\ 44\ 8 \\ \hline 3\ 20 \\ 3\ 20 \\ \hline 0 \end{array}$ vs $\dfrac{\cancel{48}\,3}{\cancel{64}\,4} = \dfrac{3}{4}$ $\begin{array}{r} .75 \\ 4\overline{)3.00} \\ 2\ 8 \\ \hline 20 \\ 20 \\ \hline 0 \end{array}$

UNIT 10

FRACTION - DECIMAL EQUIVALENT CHART

INTRODUCTION

The Fraction-Decimal Equivalent Chart is a convenient reference source found in handbooks and on wall charts. Machinists use this type of chart so frequently that they can recite most of the conversions from memory. But, initially, the machinist will need to develop skills in converting fractions, decimals, and millimeters to ensure that accurate conversion numbers are obtained when the chart is used.

OBJECTIVE:

After completing this unit the student will be able to:

- Determine the exact decimal equivalent for shop fractions
- Determine the nearest common Shop Fraction when given a decimal numeral

CHANGE FRACTIONS TO EQUIVALENT DECIMALS

The decimal equivalent for a given fraction can be found from a Decimal Equivalent Chart (or Table). Such charts are commonly used in the shop, and consist of **SHOP FRACTIONS**.

MEMORIZE THE BASIC SHOP FRACTION—DECIMAL EQUIVALENTS

$\frac{1}{2}$ = .500 $\qquad$ $\frac{1}{16}$ = .0625

$\frac{1}{4}$ = .250 $\qquad$ $\frac{1}{32}$ = .03125

$\frac{1}{8}$ = .125 $\qquad$ $\frac{1}{64}$ = .015625

CHANGE DECIMALS TO NEAREST FRACTION

💡 To change a decimal to the nearest **SHOP FRACTION**:

    Step 1   Express the given decimal using the required number of places.

    Step 2   Locate the nearest decimal on the chart (closest in value).

    Step 3   Read the fractional equivalent for this number.

EXAMPLE 1: (Exact Value)

    Express .875 as a fraction.

    Locate .875 in the chart.

    Read the fractional value which is $\frac{7}{8}$ (.875 = $\frac{7}{8}$), or

EXAMPLE 2: (Approximate Value)

    Express .646 as a fraction

    Locate .646 on the chart. It is found between .640625 and .65625.

    .640 and .656 have been underlined to show the same number of decimal places as the original number (.646).

    .646 differs from .640 by 6 thousandths (.006)

    .646 differs from .656 by 10 thousandths (.010).

The decimal having the least difference from the required number is .640. Therefore, we use the equivalent fraction for .640 as the nearest common fraction for .646.

$$\text{or } (.646 \approx \frac{41}{64})$$

FRACTION—DECIMAL EQUIVALENT CHART

Fraction				Decimal	Millimeters	Fraction				Decimal	Millimeters
8ths	16ths	32nds	64ths			8ths	16ths	32nds	64ths		
			1	0.015625	0.396875				33	0.515625	13.096875
		1		.031250	0.793750			17		.531250	13.493750
			3	.046875	1.190625				35	.546875	13.890625
	1			.062500	1.587500		9			.562500	14.287500
			5	.078125	1.984375				37	.578125	14.684375
		3		0.093750	2.381250			19		0.593750	15.081250
			7	.109375	2.778125				39	.609375	15.478125
1				.125000	3.175000	5				.625000	15.875000
			9	.140625	3.571875				41	.640625	16.271875
		5		.156250	3.968750			21		.656250	16.668750
			11	0.171875	4.365625				43	0.671875	17.065625
	3			.187500	4.762500		11			.687500	17.462500
			13	.203125	5.159375				45	.703125	17.859375
		7		.218750	5.556250			23		.718750	18.256250
			15	.234375	5.953125				47	.734375	18.653125
2				0.250000	6.350000	6				0.750000	19.050000
			17	.265625	6.746875				49	.765625	19.446875
		9		.281250	7.143750			25		.781250	19.843750
			19	.296875	7.540625				51	.796875	20.240625
	5			.312500	7.937500		13			.812500	20.637500
			21	0.328125	8.334375				53	0.828125	21.034375
		11		.343750	8.731250			27		.843750	21.431250
			23	.359375	9.128125				55	.859375	21.828125
3				.375000	9.525000	7				.875000	22.225000
			25	.390625	9.921875				57	.890625	22.621875
		13		0.406250	10.318750			29		0.906250	23.018750
			27	.421875	10.715625				59	.921875	23.415625
	7			.437500	11.112500		15			.937500	23.812500
			29	.453125	11.509375				61	.953125	24.209375
		15		.468750	11.906250			31		.968750	24.606250
			31	.484375	12.303125				63	.984375	25.003125
4				.500000	12.700000	8	16	32	64	1.000000	25.400000

NOTE: Table is exact; all figures beyond the six places given are zeros.
BASIS: 1 inch = 25.4 millimeters provides exact six place values.

OBSERVE: The metric equivalent for common fractions is shown for information purposes. This will be explained in Unit 19 on **METRICS**.

UNIT 11

ADDITION AND SUBTRACTION OF DECIMALS

INTRODUCTION

Using the height gage for layout requires adding and subtracting of decimals from point to point on the layout. The machinist should keep a running calculation of all points so that the accuracy of the layout can be checked against the blueprint before machining begins. Remember always to check the accuracy of calculations before beginning work. A part with precisely machined surfaces may become high priced scrap metal if the work was based on inaccurate calculations.

OBJECTIVES:

After completing this unit the student will be able to:

    Add decimals

    Subtract decimals

To add or subtract decimals it is necessary to preserve **PLACE VALUE** of the individual digits. This combining process requires that the tenths place of one number be grouped with the tenths place of the other number(s). In a similar manner the hundredths are all grouped together, etc.

CONSIDER: The problem of adding 3.7621 and 54.137. These numbers are arranged in columns with the decimal points serving as your guide.

Tens	Ones	Tenths	Hundredths	Thousandths	Ten Thousandths
	3	7	6	2	1
+5	4	1	3	7	
5	7	8	9	9	1

This arrangement automatically places all corresponding place values in vertical columns, and may now be added or subtracted. The **sum is 57.8991.**

🔎 It is sometimes useful to add zeros to the given numbers so that all will have the same number of decimal places.

EXAMPLE 1: (Addition)

Add: 233.3,   3.647,   37.42   and   .436

```
   243.300        (two zeros added)
     3.647
    37.420        (one zero added)
+     .436
─────────
   284.803
```

EXAMPLE 2: (Subtraction)

Subtract: 15.432   from   39.37

```
   39.370        (one zero added)
 − 15.432
─────────
   23.938
```

CONCEPT APPLICATIONS

Set 1

Add the following:

(1) .025 + .0999 = _____

(2) .0003 + .070 + .005 = _____

(3) .8507 + .010 = _____

(4) 3.087 + .062 + .001 + 456 = _____

(5) .0001 + .010 + .001 = _____

(6) 2.503 + .056 + 3.1 = _____

(7) .001 + 2.078 + .01 = _____

(8) 3.678 +· .50 + 813 + .0003 = _____

(9) 7.010 + .100 + .0001 = _____

(10) .625 + .125 + .3125 + .0625 = _____

Set 2

Subtract the following:

(1) 8948.000 − .8940 = _____

(2) 1.000 − .490 = _____

(3) 7.000 − .600 = _____

(4) 2.310 − .0075 = _____

(5) 2.125 − 2.105 = _____

(6) 1.0025 − .9998 = _____

(7) 125.000 − .125 = _____

(8) 3.0625 − 2.573 = _____

(9) 42.630 − 18.735 = _____

(10) .001 − .0001 = _____

Set 3

(1) Find:

X = _____
Y = _____
Z = _____
W = _____

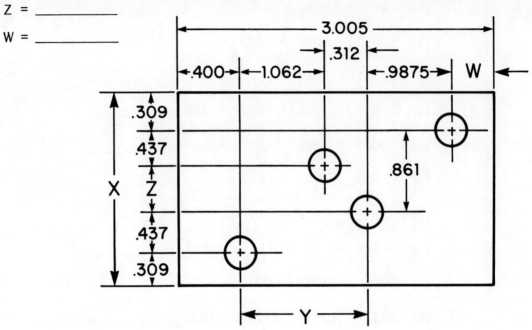

(2) Find:

A = _____
B = _____
C = _____

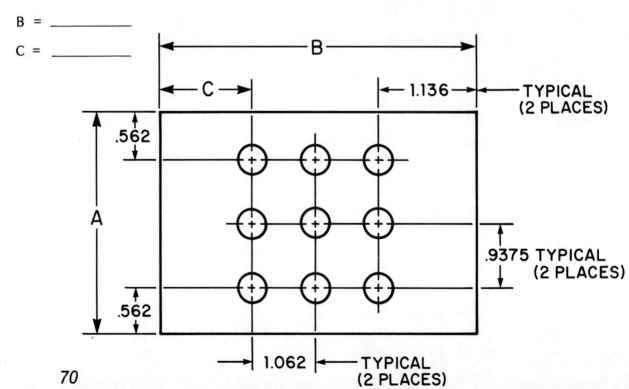

(3) Find A = _____ B = _____ C = _____

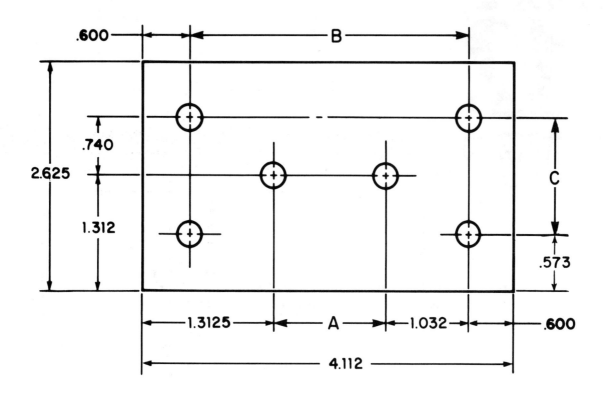

(4) Find A = _____ B = _____ C = _____ D = _____

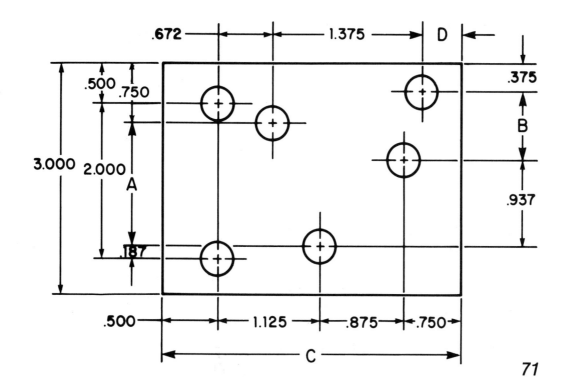

71

UNIT 12

MULTIPLICATION OF DECIMALS

INTRODUCTION

Multiplication of decimals requires use of simple math skills and application of common sense. Calculating cutting speeds and feeds and converting metric measurements to inch measurements are typical applications requiring these. Care must be exercised in placing the decimal to ensure that it is in the right location and not off by a factor of x10 or even x100.

OBJECTIVE:

After completing this unit the student will be able to:

- Multiply decimals

Mulitplication of decimals is very similar to multiplication of whole numbers. The only difference is that when working with decimals allowances must be made for the different **PLACE VALUES** assigned to the digits.

TO PROPERLY PLACE THE DECIMAL POINT:

Step 1 Determine the **total** number of digits in the problem which are found to the **right** of the decimal points.

Step 2 Place the decimal point in the answer (product) such that it has the same total number of decimal places as determined in Step 1 above.

37/3600 (upside down)

225

3/ 1 1/2

2.625

$A^2 + B^2 = C^2$

35° 38' 41"

.75   1.300/323
       1.062

$$\begin{array}{r} 29 \\ \times 3 \\ \hline 87 \end{array}$$

$$\begin{array}{r} 31.5 \\ 29\overline{)144} \\ \underline{87} \\ 374 \\ \underline{29} \\ 84 \\ \underline{48} \\ 36 \end{array}$$

$$\begin{array}{r} 29 \\ \times 2 \\ \hline 48 \end{array}$$

$$\begin{array}{r} 29 \\ \times 4 \\ \hline 1\,6 \end{array}$$

$$\begin{array}{r} 10 \\ \times 039370 \\ \hline \end{array}$$

$$\begin{array}{r} 36 \\ \times 25.4 \\ \hline \end{array}$$

EXAMPLE 1:

21.5	1	digit to the right of decimal point
x 5	+ 0	digits to the right
107.5	1	digit = **total** places required for your answer (product)

or

EXAMPLE 2:

31.5	1	digit to the right
x .27	+ 2	digits to the right
2205	3	digits = total required for the product
630		
8.505		

stated simply:

EXAMPLE 3:

21.238	3	digits
x 1.3	+ 1	digit
63714	4	digits
21238		
27.6094		

73

CONCEPT APPLICATIONS

Multiply

(1) 3.25 x .05 = _____

(2) .675 x 4.275 = _____

(3) The circumference of a circle is equal to 3.14 times the diameter. Find the circumferences of the following circles with the given diameters:

    a) d = 3.673   C = _____   c) d = 12.6   C = _____

    b) d = 1.010   C = _____   d) d = .875   C = _____

(4) 4.75 x .25 = _____

(5) .675 x 4.275 = _____

(6) The area of a square is the product of its sides, s x s = A
Find the area of the following squares:

    a) s = 2.250   A = _____   c) s = 5.5   A = _____

    b) s = .4375   A = _____   d) s = .062   A = _____

(7) How much stock will it require to make 500 pieces 3.250" long? Ans = _____

(8) If the rate of assembly is averaged at 206.7 pieces per hour. How many parts can be assembled in an 8 hour shift? Ans = _____

(9) Each plastic part weighs .087 pounds. What is the weight of 20,000 parts? Ans = _____

(10) If it takes 25 tons of pressure to shear a one square inch cross section of mild steel, how many tons of pressure will it require to shear 34.837 square inches? Ans = _____

(11) There are 25.4 millimeters in one inch. Convert the following inch dimensions to millimeters:

    a) .010" = _____ mm   e) 2.3125" = _____ mm

    b) .153" = _____ mm   f) 1.3266" = _____ mm

    c) .375" = _____ mm   g) 4.275" = _____ mm

    d) .979" = _____ mm   h) 34.972" = _____ mm

(12) If the hourly rate of a worker is $9.75 per hour and the overtime rate is 1.5, what is the amount of overtime pay for 50 hours based on a 40 hour week? Ans = $_____

(13) A print calls for 7 equally spaced holes 1.3125" apart center to center. What is the distance between centers of the first and last hole?

Ans = _____

(14) If a compression spring requires 37.2 lbs to compress .10 inch, how many pounds would it exert at .625 inches?   Ans = _____ lbs.

(15) A pressure pad in a die set contained 16 springs. If each spring requires 97.6 lbs. to compress it .10 of an inch, what would be the force on the pad if the pad were compressed .750 of an inch?   Ans = _____

UNIT 13
DIVISION OF DECIMALS

INTRODUCTION

Machinists are often required to calculate the number of threads of a given pitch there are on a specific length of threading. Determining cutting speeds and feeds requires the use of math skills, such as division, that are just as fundamental for the machinist as is the ability to perform machining operations within tolerance.

OBJECTIVES:

After completing this unit the student will be able to:

- Divide decimals by whole numbers
- Divide whole numbers by decimals
- Divide decimals by decimals
- Check division answers

DEFINITIONS:

  DIVISOR   A number used to divide a dividend.

  QUOTIENT   The number which results from dividing. (The answer)

  DIVIDEND   The number being divided.

  REMAINDER   The number left over after any step in the division process

DIVIDING DECIMALS BY WHOLE NUMBERS

EXAMPLE 1: (with zero remainder)

```
        .26  ←_____
  ___→ 5)1.30  ←_____
        1 0
        ───
         30
         30
         ──
          0  ←_____
```

divisor is 5
quotient is .26
dividend is 1.30 or 1.3
remainder is zero (0)

Label the blanks

CHECK:

(dividend) = (divisor × quotient) + (remainder)

(1.30) = (5 × .26) + (0)

EXAMPLE 2: (with remainder)

```
      2.669
   5)13.347
     10
     ──
      3 3
      3 0
      ───
        34
        30
        ──
         47
         45
         ──
          2
```

divisor is _____
quotient is _____
dividend is _____
remainder is _____*

*(in the thousandths place)

CHECK:

(dividend) = (divisor × quotient) + (remainder)

(13.347) = (5 × 2.669) + (.002)

Examples 1 and 2 above illustrate the rule for dividing a decimal by a whole number.

RULE:

Step 1 Divide as you would with whole numbers

Step 2 Place the decimal point straight up into the quotient

Locate the decimal point in this quotient:

```
      26 3
   8)210.4
```

26.3 is correct

77

CHECK:

$$\begin{array}{r} 26.3 \\ \times 8 \\ \hline 210.4 \end{array}$$

DIVIDING BY A DECIMAL

A second type of division involves dividing a **whole number** or **decimal** by a **decimal**.

RULE: If the **divisor** is a decimal --- then

Step 1   Rename the divisor as a whole number by moving the decimal point to the right as many decimal places as are given in the divisor.

Step 2   Move the decimal point an equal number of places in the dividend.

The result of this relocation process gives us a problem identical to those in examples 1 and 2. Therefore, we proceed using the method described in examples 1 and 2.

EXAMPLE 3: ("∧" shows new location of decimal point)

$$3.6\overline{)17.28} \qquad 36\overline{)\begin{array}{r}4.8\\172.8\\144\\\hline 28\ 8\\28\ 8\\\hline 0\end{array}}$$

step 1 → ↑ ↑ ← step 2

EXAMPLE 4:

$$.005\overline{).67525} \qquad 5\overline{)675.25}$$

(3 place move)

Complete steps to show the quotient = 135.05.

Sometimes one or more ZEROS are needed in order to carry out the division.

Note the use of zero(s) in the following:

$$8\overline{)\begin{array}{r}.0009\\.0072\end{array}} \qquad .06\overline{)36.} \rightarrow 6\overline{)\begin{array}{r}600.\\3600.\end{array}} \qquad .017\overline{)81.6} \rightarrow 17\overline{)\begin{array}{r}4800\\81600\end{array}}$$

CONCEPT APPLICATIONS:

Divide problems (1) – (8) to the third decimal place (thousandths).

(1) $.3 \overline{)20}$ (5) $8 \overline{)1.}$

(2) $.07 \overline{)3.}$ (6) $16 \overline{)1.}$

(3) $.61909 \overline{)18.30408}$ (7) $32 \overline{)1.}$

(4) $25.4 \overline{)1.00000}$ (8) $64 \overline{)1.}$

(9) If it takes .067 lbs to produce a plastic part, then how many parts can be produced from a hundred pounds of plastic?

(10) If a worker produced 3672 parts in an 8 hour day, then how many parts did the worker produce per minute?

(11) What is the distance between 8 equally spaced holes if the distance between the first and eighth hole is 3.062?

(12) If the pitch of a thread is 13 threads per inch, how many threads will a threaded portion have that is 3.250" long? ($P = \frac{1}{n}$)

(13) If the pitch of a progressive die (distance the part moves between stations) is .937, how much stock will be fed into the die before a final part is produced in a 7 station die?

(14) If parts 1.76 square inches in area are produced, how many parts can be produced from a 1 inch strip 60 feet long?

UNIT 14

DECIMAL CONCEPT APPLICATIONS

OBJECTIVE:

After completing this unit the student will be able to:

- Apply the skills of multiplication and division of decimals to practical problems

This unit is designed to give the student additional practice in solving shop related problems using both multiplication and division of decimals.

CONCEPT APPLICATIONS:

(1) One kind of Sheet Metal stock has a thickness of .0375". How many sheets of metal are there in a pile 15" high?

(2) A metal strip is 39.96" long. How many pieces each 2.125" long, can be cut from it if .095" waste is allowed for each piece?

(3) How many sheets of metal 0.1875" thick can be placed in a box 4.500" deep?

(4) How many cuts would be needed to turn down 1.460" stock to 1" on a lathe, each cut being .023" deep?

(5) If an alloy is 67% copper and 33% zinc, how many pounds of each metal will there be in a casting weighing 90 pounds?
(% = Hundredths)

(6) If 936 pieces are machined by 37 men, how many pieces did each man machine?

(7) An assembly consists of 4 parts weighing 348 lbs., 361 lbs., 368 lbs., and 173 lbs. Nine of these assemblies are to be made. If these assemblies are to be delivered in the shop's truck, which has a capacity of 2 tons, how many trips will the truck have to make?
(one ton = 2000 pounds)

(8) An assembly is made up of six parts having the following weights: 7.4 lb., 348 lb., 2.21 lb., 95 lb., 16 lb., and 5.8 lb. What is the weight of the complete assembly?

(9) A production machine makes 7 parts in 2 minutes, 15 seconds. At this rate, how many parts will be made in an 8 hour day?

(10) A job that requires 1144 hrs. was divided equally among 29 men. How many hours and minutes must each man work?

(11) If a man works 8 hours a day, 5 days a week, for 50 weeks at $5.12 per hour, what does he earn?

(12) How many 7.875 inch pieces can be cut from a bar 12 feet long?

(13) A mechanic can assemble 4 machines every 6 hours. How many machines can 48 mechanics assemble in a month of 25 eight hour working days?

(14) If a milling-machine operator finishes a gear in 2.6 hours, how many eight hour days will it take him to deliver 240 gears? How many 5 day weeks is this?

(15) What is the cost of a bar of stock which weighs 118 lbs. and costs $69.55/hundred weight?

(16) A shop uses 96.8 kilowatt-hours of electricity in a 24 day month. What is the average consumption per day.

(17) A certain size of bar stock weighs 24 ounces per linear foot. How much will an 8-foot section weigh? How much will an 11 inch section weigh?

(18) If it takes 45 seconds to cut through a section of 3-inch diameter bar stock, how long will it take to cut 12 pieces, each 3 inches long, from a 48 inch length of this bar?

(19) If an automatic screw machine finishes 1 part in 37 seconds, how many parts are produced in one hour?

(20) An assembly weighing 1862.5 lbs consists of five components. What is the average weight of a component?

UNIT 15

TOLERANCES

INTRODUCTION

Tolerances provide the degree of accuracy required in making machined parts. Since it is impossible to duplicate an exact measure or dimension, the tolerance is a realistic guide that limits how far a dimension is allowed to vary from a basic dimension and still be acceptable. The acceptable tolerance range has a great influence on the cost of the final product. Dimensions with larger tolerances are easier to machine and, therefore, can be made faster and cheaper than those with closer tolerances.

OBJECTIVES:

After completing this unit the student will be able to:

- Define the terms used in shop measurement
- Determine bilateral dimensions
- Determine unilateral dimensions
- Determine maximum dimensions
- Determine minimum dimensions
- Determine tolerance from given limits

DEFINITIONS:

MEASURE

An indication of comparative size.

TOLERANCES

A determination of an acceptable range of measurement.

LIMITS

The largest (maximum) and smallest (minimum) acceptable measure for a range of measurement.

BASIC DIMENSION

The exact measure or the desired measure between two specific points.

BILATERAL TOLERANCE

A measure which is added to and subtracted from a basic dimension to establish maximum and minimum limits.

UNILATERAL TOLERANCE

A measure which is either added to or subtracted from a basic dimension, but not both.

MAXIMUM DIMENSION (LIMIT)

A measure used to identify the largest acceptable range.

MINIMUM DIMENSION (LIMIT)

A measure used to identify the smallest acceptable range.

NOTE:

Tolerances for METRIC SYSTEM are covered in Unit 19 Metric Section.

TOLERANCE:

The success of a machine operator depends on the ability to work with measures. For the machine operator, measures are usually given in fractional or decimal form. Decimal measures are the most common, such as 2.312. (Expressed in inches unless otherwise stated). **It is impossible to duplicate an exact measure.** The degree of accuracy depends on the instrument used for measuring, the machine which is to produce the product and the person responsible for using both. However, it is generally possible to produce a product whose dimensions are within an acceptable range. The amount that a measure can vary and still be acceptable is referred to as the **"tolerance"** for the basic dimension.

83

BILATERAL TOLERANCE.

EXAMPLE: Given the following:

$$2.132 \pm .005$$

Basic Dimension ⟶ (2.132)
Bilateral Tolerance ⟶ (± .005)

This means that the acceptable measure for the product must be within a specific range. The range is determined by finding its Limits. **The UPPER LIMIT (maximum) is equal to the sum of the Basic Dimension (2.132) and the Positive Tolerance (+.005).** Or, 2.132 + .005 = 2.137.

The LOWER LIMIT (minimum) is equal to the difference between the Basic Dimension (2.132) and the Negative Tolerance (−.005). Or, 2.132 − .005 = 2.127.

The actual measure of an acceptable product must fall within 2.127 and 2.137. The Limits give us a range of tolerance of .010. **The RANGE is equal to the difference between the upper and lower limits.** Or, 2.137 − 2.127 = .010.

UNILATERAL TOLERANCE.

EXAMPLE: Given the following:

$$3.446 \begin{array}{l} +.005 \\ -.000 \end{array}$$

Basic Dimension ⟶
Unilateral Tolerance ⟶

This means that the range is to be found using the Basic Dimension as the Lower Limit. **The UPPER LIMIT is equal to the sum of the Basic Dimension (3.446) and the Positive Tolerance (+.005).** Or, 3.446 + .005 = 3.451 which is the Upper Limit.

The LOWER LIMIT, in this case, is the Basic Dimension, 3.446 − .000.

In this case our RANGE of acceptable measure is identified by the limits of 3.451 and 3.446. Therefore, our Range of tolerance is:

$$3.451 - 3.446 = .005$$

STUDY THE FOLLOWING:

EXAMPLE 1:

Given the dimension   2.312   ± .005

The dimension is bilateral
The maximum dimension is = 2.317
The minimum dimension is = 2.307
The tolerance is   = .010
The basic dimension is   = 2.312

EXAMPLE 2:

Given the dimension   .875 $\begin{matrix} + .000 \\ - .010 \end{matrix}$

The dimension is Unilateral
The maximum dimension is = .875
The minimum dimension is = .865
The tolerance is .010
The basic dimension = .875

EXAMPLE 3:

Given the dimension $1\frac{5}{8}$ ± $\frac{1}{64}$

The dimension is bilateral
The maximum dimension is = $1\frac{41}{64}$

The minimum dimension is = $1\frac{39}{64}$

The tolerance is $\frac{1}{32}$

The basic dimension is = $1\frac{5}{8}$

CONCEPT APPLICATIONS

Complete the chart for the following dimensions:

    Type of dimension Bilateral (B), Unilateral (U)
    Basic dimension (Ba)
    Maximum dimension (Max)
    Minimum dimension (Min)
    Tolerance (T)

		B	U	Ba.	Max	Min	T
(1)	2.312 ± .005	√		2.312	2.317	2.307	.010
(2)	.875 +.000/−.010		√	.875	.875	.865	.010
(3)	.062 ± .003						
(4)	.1875 +.0003/−.0000						
(5)	.3125 ± .0001						
(6)	.4375 ± .0015						
(7)	.5000 +.0000/−.0001						
(8)	.5625 ± .0025						
(9)	.625 ± .004						
(10)	.675 +.001/−.000						
(11)	.750 ± .006						
(12)	.8125 ± .0007						
(13)	1.875 ± .001						

	B	U	Ba.	Max	Min	T
(14) 5.9375 ± .0004						
(15) 1.250 +.0003/−.0000						
(16) 3.0312 ± .0005						
(17) 1.1093 +.0000/−.0001						
(18) 1.2031 ± .0004						
(19) 2.718 +.0001/−.0000						
(20) 1.0000 ± .0004						
(21) .718 +.000/−.005						
(22) .281 ± .002						
(23) 1.2054 ± .0005						
(24) .749 +.000/−.004						
(25) .8748 +.0000/−.0005						
(26) 9.375 ± .010						
(27) .500 +.030/−.000						
(28) .252 ± .001						

UNIT 16
PERCENT

INTRODUCTION

Percent is a form of comparision with 100 fixed or understood as the denominator. Percent of material thickness, percent of cutting clearance, percent of shrinkage and percent of scrap are examples. Percent expresses a ratio that shows how much a measure varies from a unit.

OBJECTIVES:

After completing this unit the student will be able to:

- change decimals to percents
- change percents to decimals
- change fractions to percents
- change percents to fractions
- solve percentage problems

UNDERSTANDING PERCENT

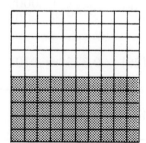

In the square at the right, 50 of the 100 squares have been shaded. This ratio of 50 out of 100 is expressed as $\frac{50}{100}$ or 50%. (50% is read as "50 percent".)

 Percent means Per Hundred:

$$1 \text{ percent} = \frac{1}{100} = .01 \text{ etc.}$$

EXAMPLES:

Write a percent for each of the following:

a. 75 out of 100 =   b. $\frac{18}{100}$ =

SOLUTIONS:

a. $\frac{75}{100}$ or 75%.
b. 18 out of 100 or 18%.

Shade 75% of each of the following:

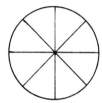 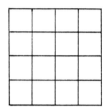 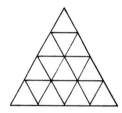

Parts to be shaded:

.75 x 8 = _____   .75 x ____ = _____   .75 x ____ = _____

CHANGE DECIMAL TO PERCENT

RULE — To change a decimal number to a percent, **multiply the decimal by 100** and **add the percent symbol**.

EXAMPLES:

a. Change .625 to a percent.
b. Change .25 to a percent.
c. Change $.87\frac{1}{2}$ to a percent.

SOLUTIONS:

a. .625 x 100 = 62.5% or $62\frac{1}{2}$%.
b. .25 x 100 = 25%
c. $.87\frac{1}{2}$ x 100 = $87\frac{1}{2}$% or 87.5%

CHANGE PERCENT TO DECIMAL

RULE — To change a percent to a decimal, **move the decimal point two places to the left** and **drop the percent symbol**. (insert zeros as needed).

EXAMPLES:

a. 42% = .42   d. 1% = .01
b. 5% = .05   e. $16\frac{2}{3}$% = $.16\frac{2}{3}$
c. 50% = .50   f. $87\frac{1}{2}$% = $.87\frac{1}{2}$ or .875

89

CHANGE FRACTION TO PERCENT

RULE – To change a fraction to a percent, **change the fraction to a decimal and multiply the decimal by 100.** Then **add the percent (%) symbol.**

EXAMPLE 1:   Write a percent for $\frac{5}{8}$.

$$\frac{5}{8} \rightarrow 8 \overline{\smash{)}\begin{array}{l}.62\frac{4}{8}\\5.00\\\underline{4\ 8}\\\ \ \ 20\\\ \ \ \underline{16}\\\ \ \ \ \ 4\end{array}}$$

$.62\frac{4}{8} = .62\frac{1}{2}$ or $.625$

$.625 \times 100 = 62.5\%$ or $62\frac{1}{2}\%$

EXAMPLE 2:   Write a percent for $\frac{9}{12}$

$$\frac{9}{12} = \frac{3}{4} \rightarrow 4\overline{\smash{)}\begin{array}{l}.75\\3.00\\\underline{2\ 8}\\\ \ \ 20\\\ \ \ \underline{20}\end{array}}$$

$.75 \times 100 = 75\%$

CHANGE PERCENT TO FRACTION

RULE – To change a percent to a fraction, **write the percent over 100**, **drop the percent symbol (%)** and **reduce the fraction to lowest terms.**

EXAMPLE 1:   Change 75% to a fraction

$$75\% = \frac{75\%}{100} \rightarrow \frac{\cancel{75}^{3}}{\cancel{100}_{4}} = \frac{3}{4}$$

or $75\% = \frac{3}{4}$

EXAMPLE 2:   Change $12\frac{1}{2}\%$ to a fraction

$$12\frac{1}{2}\% = \frac{12\frac{1}{2}\%}{100} \rightarrow \frac{12\frac{1}{2}}{100} \rightarrow \frac{25}{2} \div \frac{100}{1}$$

$$\frac{\cancel{25}^{1}}{2} \times \frac{1}{\cancel{100}_{4}} = \frac{1}{8}$$

or $12\frac{1}{2}\% = \frac{1}{8}$

THERE ARE THREE TYPES OF PERCENT PROBLEMS TO BE CONSIDERED

Type I   Finding the percent of a given number.

Type II Finding what percent one number is of another number.

Type III Finding a number when a percent of it is known.

TYPE 1 — FINDING THE PERCENT (%) OF A NUMBER.

RULE —   Change the % to a decimal. Then multiply by the number.

EXAMPLE:

Find — 72% of 250 = ___?___

$$72\% = .72$$

```
    250
  x .72
    500
   1750
  180.00
```

TYPE 2 — FINDING WHAT PERCENT (%) ONE NUMBER IS OF ANOTHER

RULE —   Write a fraction for the %. Then divide the numerator by the denominator.

EXAMPLE 1: ___?___ % of 20 = 17 ⟶ means $\frac{17}{20}$ = ___?___ %

___?___ % of 20 = 17

$\frac{17}{20}$ = 17 ÷ 20

```
       .85
  20 ) 17.00
       16 0
        1 00
        1 00
```

.85 = 85% or
<u>85%</u> of 20 = 17

91

EXAMPLE 2: __?__ % of 63 = 18 ⟶ means $\frac{18}{63}$ = __?__ %

__?__ % of 63 = 18

$\frac{18}{63}$ = 18 ÷ 63

$63 \overline{)18.00}$ or

since $\frac{\cancel{18}^{\,2}}{\cancel{63}_{\,7}} = \frac{2}{7}$

$7 \overline{)2.00}.28\frac{4}{7}$
$\underline{1\,4}$
$\,60$
$\,\underline{56}$
4

$.28\frac{4}{7}$ = $28\frac{4}{7}$% or

$\underline{28\frac{4}{7}\% \text{ of } 63 = 18}$

TYPE 3 — FINDING A NUMBER WHEN A PERCENT (%) OF IT IS KNOWN

💡 *RULE —* Write a decimal for the %. Then divide the number by the decimal.

EXAMPLE 1:

12% of __?__ = 15

12% = .12

$.12 \overline{)15.00}125.$
$\underline{12}$
$\,3\,0$
$\,\underline{2\,4}$
60
$\underline{60}$

$\underline{12\% \text{ of } 125 = 15}$

EXAMPLE 2:

$6\frac{1}{4}$% of __?__ = 25

$6\frac{1}{4}$% = .0625

$.0625 \overline{)25.0000}400.$
$\underline{25\,00}$
$\,0$
$\,\underline{0}$
$\,0$
$\,\underline{0}$

$\underline{6\frac{1}{4}\% \text{ of } 400 = 25}$

NOTE:

Equivalent forms of numbers may be used in all % problems:

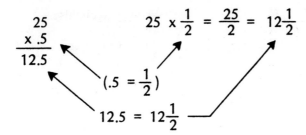

$$25 \times \frac{1}{2} = \frac{25}{2} = 12\frac{1}{2}$$

$$\begin{array}{r} 25 \\ \times .5 \\ \hline 12.5 \end{array}$$

$$(.5 = \frac{1}{2})$$

$$12.5 = 12\frac{1}{2}$$

50% of 25 = <u>12.5</u> or $12\frac{1}{2}$

CONCEPT APPLICATIONS

(1) $\dfrac{92}{368}$ = _____ %

(2) $\dfrac{63}{224}$ = _____ %

Write as decimals:

(3) 121.03% = _____

(4) .7% = _____

(5) What % of 64 is 10.880?

(6) 323 is what % of 475?

(7) Find 48% of 325.

(8) Find $\dfrac{3}{4}$% of 60.

(9) What is 13.3% of 842?

(10) A sheet of metal .130" is welded to a sheet .096" thick. The weld nugget is .0904" thick. What % of the thickness of both sheets is the nugget?

(11) A nugget thickness of 47% of the combined thickness of .065" and .053" sheets of metal. What is the nugget thickness?

(12) Change $\dfrac{7}{8}$ to percent.

(13) A shaft revolves at 245 R.P.M. It is necessary to reduce this speed by 20%. What is the speed after the reduction?

(14) On a shop run of 350 castings, 12 were defective and 8 were spoiled in machining. What % were defective? What % were spoiled? Round off to one decimal place.

(15) If the horsepower of an engine is increased 5% to 525 h.p., what was the original horsepower?

UNIT 17

SIGNED NUMBERS

INTRODUCTION

Signed numbers are simply numbers which have been given a direction that is positive or negative relative to a beginning point or origin. The movement of the cutting tool on a CNC machine is stated in positive and negative directions from an origin. Modern dimensioning techniques use datum lines and points to establish positive or negative direction of measurements from a base datum line.

OBJECTIVES:

After completing this unit the student will be able to:

- Identify and compare values of signed numbers on a number line using absolute values

- Add, subtract, multiply and divide using signed numbers

THE NUMBER LINE:

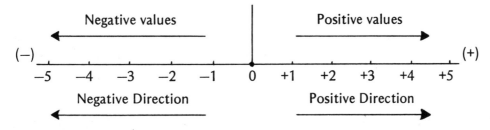

The number line **is a graphic model of the set of all real or signed numbers.** It is used to relate the value of any given number to that of zero. The basic centigrade thermometer has a number line that determines the number of degrees above (+) or below (−) zero. Zero has no sign.

ABSOLUTE VALUES:

The absolute value of a signed number is always the positive value of the number or the number without a sign. Therefore, the absolute value of +10 = 10. (Read — the absolute value of positive ten equals 10). Likewise, the absolute value of −10 = 10.

NOTE:

The symbol "| |" is used to represent the absolute value of the number(s) it contains.

EXAMPLE 1:

|+10| = 10 and
|−10| = 10, therefore
by substitution, |+10| = |−10|
since 10 = 10.

A number without a sign is considered positive. Absolute value is generally related to an actual distance or length.

EXAMPLE 2:

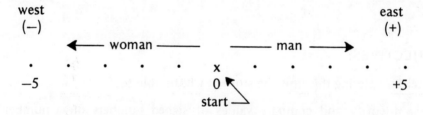

If a man and woman walk in opposite directions for a distance of 5 miles each, which person walked the greatest distance? It should be obvious that they travelled the same distance (5 miles = 5 miles) regardless of the direction they went.

Therefore, we may conclude:

|−5| miles = |+5| miles or

5 miles = 5 miles

96

RULES FOR SIGNED NUMBER OPERATIONS:

ADDITION

RULE 1   To add two or more signed numbers having the same signs (or like signs):

   Step 1   Add the absolute values of the numbers, and

   Step 2   Bring down their common sign.

EXAMPLES:

```
      + 4                                    − 9
     + 11                                    − 7
      15   ←────── Step 1 ──────→            16
    + 15   ←───── Step 2 (+ or −) ──→      − 16
```

RULE 2   To add two signed numbers having unlike (or opposite) signs:

   Step 1   Determine the **difference** between their absolute values.

   Step 2   Bring down the sign of the number having the **greatest absolute** value.

EXAMPLES:

```
  −12          −8          −4          +20
  + 6          + 9         + 6         −35
  ────         ────        ────        ────
  − 6          +1          +2          −15
```

RULE 3   To add 3 or more signed numbers having a combination of signs:

   Step 1   Find the sum of the positive numbers.

   Step 2   Find the sum of the negative numbers.

   Step 3   Find the difference in the absolute values of those sums (totals).

   Step 4   Bring down the sign of the number having the greatest absolute value.

EXAMPLES:

```
   + 4              − 25
   −12              + 37
   − 8              − 16
   + 7              + 43
   ────             ────
   + 11   (+) Total + 80
   − 20   (−) Total − 41
   ────   Difference ────
   − 9              + 39
```

CONCEPT APPLICATIONS

Add the following:

(1) 7
    8

(2) + 7
    + 8

(3) − 7
    − 8

(4) + 7
    − 8

(5) − 8
    + 9

(6) − 8
    − 9

(7) − 12 + (+ 8) = _____

(8) 13 + (+ 5) = _____

(9) − 6 + 8 + (− 3) + (− 4) = _____

(10) (− 3) + (− 4) + 6 + (− 5) + (− 4) = _____

(11) (− 9) + (− 5) + (− 4) + (− 3) = _____

(12) + 7 + 6 + (− 6) + (− 7) + (− 3) + 3 = _____

SUBTRACTION

RULE 1 To subtract signed numbers:

Step 1 Change the sign of the subtrahend.

Step 2 Proceed as in addition. (Using **ADDITION** Rules 1 and 2).

$$\begin{array}{c} +12 \\ \oplus \; \underline{-2} \\ +14 \end{array} \qquad \begin{array}{c} +12 \\ \ominus \; \underline{+2} \\ +10 \end{array} \qquad \begin{array}{c} -12 \\ \ominus \; \underline{+2} \\ -14 \end{array} \qquad \begin{array}{c} -12 \\ \oplus \; \underline{-2} \\ -10 \end{array}$$

NOTE:

This subtraction rule is often stated: To subtract a number, add its opposite.

EXAMPLES:

$$\begin{array}{c} -15 \\ \underline{+15} \end{array} \longrightarrow \begin{array}{c} -15 \\ \underline{-15} \\ -30 \end{array} \quad \text{or} \quad \ominus \begin{array}{c} -15 \\ \underline{+15} \\ -30 \end{array}$$

$$\begin{array}{c} -4 \\ \underline{-10} \end{array} \longrightarrow \begin{array}{c} -4 \\ \underline{+10} \\ +6 \end{array} \quad \text{or} \quad \oplus \begin{array}{c} -4 \\ \underline{-10} \\ +6 \end{array}$$

$$\begin{array}{c} +12 \\ \underline{+15} \end{array} \longrightarrow \begin{array}{c} +12 \\ \underline{-15} \\ -3 \end{array} \quad \text{or} \quad \ominus \begin{array}{c} +12 \\ \underline{+15} \\ -3 \end{array}$$

NOTE:

The form on the right with the "O" reduces the chance of making an error when changing signs.

CONCEPT APPLICATION

Subtract:

(1)   9
     −5

(2)   12
     −7

(3)   15
     −9

(4)   16
     −7

(5)   8
     +2

(6)   7
     +4

(7)   7
     −4

(8)   −7
     +4

(9)   −7
     −4

(10)   4
     +4

(11) $4 - (-4) =$ _____

(12) $-4 - (+4) =$ _____

(13) $-4 - (-4) =$ _____

(14) $11 - (+6) =$ _____

(15) $-11 - (+6) =$ _____

(16) $-11 - (-6) =$ _____

(17) $+11 - (-6) =$ _____

(18) $5 - (-5) =$ _____

(19) $-5 - (-5) =$ _____

(20) $-5 - (+5) =$ _____

MULTIPLICATION

RULE 1: To multiply two signed numbers having **like** signs:

Step 1 Multiply the absolute values ⟶ (product)

Step 2 Precede this product with a positive (+) sign.

EXAMPLES:

$$\begin{array}{cccc} +4 & +6 & -4 & -6 \\ \times +7 & \times +3 & \text{or} \quad \times -7 & \times -3 \\ \hline +28 & +18 & +28 & +18 \end{array}$$

(compare products carefully):

$(+4) \times (+7)$ is equal to $(-4) \times (-7)$ since both are equal to **+28**.

RULE 2: To multiply two signed numbers having **unlike** signs:

Step 1 Multiply the absolute values ⟶ (product)

Step 2 Precede this product with a negative (−) sign.

EXAMPLES:

$$\begin{array}{cccc} -4 & +6 & +4 & -6 \\ \times +7 & \times -3 & \times -7 & \times +3 \\ \hline -28 & -18 & -28 & -18 \end{array}$$

RULE 3 To multiply three or more signed numbers (factors):

Step 1 Multiply their absolute values ⟶ (product)

Step 2 The product is positive if an even number of negative factors are included.

Step 3 The product is negative if an odd number of negative factors are included.

NOTE:
Multiplication of numbers may be indicated as follows in symbols:

OBSERVE:

General forms $(a \cdot b) = (a) \times (b) = (a)(b) = (ab) = ab$

(negative 5) x (positive 6) =
$(-5) \times (+6) =$
$(-5) \quad (+6) =$ drop "x"
$(-5) \cdot (+6) =$ $\cdot = (\times)$
$-5 \cdot +6 =$ drop ( )
$-5 \cdot 6 =$ simplest form

EXAMPLE 1:

$(+2)(-2)(-2)(-2) = -16$ *Note:* (odd) negative sign → $(-)$

EXAMPLE 2:

$(-1)(-1)(+2)(+3)(-2)(-3) = +36$ *Note:* (even) negative signs → $(+)$

CONCEPT APPLICATION

Multiply:

(1) -2
       -3

(2) -8
       6

(3) 3
       -5

(4) 7
       -9

(5) -9
       -8

(6) $(3) \times (-2) \times (-1) = $ _____

(7) $(+3) \times (+4) \times (+2) = $ _____

(8) $(+5) \times (+6) \times (+3) = $ _____

(9) $(+4) \times (-3) \times (-2) = $ _____

(10) $(-7) \times (-3) \times (+3) = $ _____

DIVISION

RULE 1 To divide signed numbers having **like** signs:

    Step 1 Divide the absolute values $\longrightarrow$ (quotient)

    Step 2 Precede this quotient with a positive (+) sign.

EXAMPLES:

$$(+21) \div (+7) = +3$$

$$(-21) \div (-7) = +3$$

$$\frac{-40}{-8} = +5$$

$$\frac{+40}{+8} = +5$$

RULE 2 To divide signed numbers having **unlike** signs:

    Step 1 Divide the absolute values $\longrightarrow$ (quotient)

    Step 2 Precede this quotient with a negative (−) sign.

EXAMPLES:

$$(-36) \div (+18) = -2$$

$$(+36) \div (-18) = -2$$

$$\frac{-12}{+12} = -1$$

$$\frac{-4}{+16} = -\frac{1}{4}$$

CONCEPT APPLICATION

Divide:

(1) $14 \div -7 =$

(2) $-3 \div -1 =$

(3) $-5 \div 5 =$

(4) $4.28 \div -2 =$

(5) $6.2 \div -3.1 =$

(6) $\dfrac{-16}{8} =$

(7) $\dfrac{12}{-4} =$

(8) $\dfrac{-11}{2} \div \dfrac{2}{3} =$

(9) $\dfrac{-5}{8} \div \dfrac{-1}{8} =$

(10) $-.456 \div 2.4 =$

(11) $46 \div 3.5 =$

(12) $24 \div -2.46 =$

(13) $\dfrac{-7}{8} \div .625 =$

(14) $.3125 \div -2 =$

UNIT 18
POWERS AND ROOTS

INTRODUCTION

A power of a number is a way of expressing multiplication of a base number by itself. Powers are useful to find areas of circles or volumes of solids. The reverse process, determining a root, allows us to find dimensions if the area or volume is known.

OBJECTIVES:

After completing this unit the student will be able to:

- Find the square and square root of a whole number or fraction
- Raise a number to any power
- Simplify expressions using powers and roots

DEFINITIONS:

 FACTOR

    A number being multiplied by any other number(s):

REVIEW:

$3 \cdot \underline{?} = 12$ Use the raised dot for multiplication

    ↑—factor   ↑—product

The missing **factor** __?__ is 4

FACTORIZATION

A set of two or more factors which are equal to a particular product: (2 · 6) = (3 · 4) = (2 · 2 · 3). Each of these ( ) is a **factorization** of the number 12.

NOTE:

All numbers have a factor of 1, although it is generally not written as part of the factorization:

$$(2 \cdot 3) = (1 \cdot 2 \cdot 3) = 6$$

EXPONENT

A number that tells how many times a specific number is to be used as a factor. The exponent is a smaller sized number written to the upper right of the factor.

$$5^3 = 5 \cdot 5 \cdot 5 = \underline{?}$$

$$3^4 = 3 \cdot 3 \cdot 3 \cdot 3 = \underline{?}$$

POWER

A product which contains a given factor an indicated number of times:

EXAMPLE 1: 27 is the third power of 3, or $3^3 = 3 \cdot 3 \cdot 3 = 27$

EXAMPLE 2: 16 is the fourth power of 2, or $2^4 = 2 \cdot 2 \cdot 2 \cdot 2 = 16$

NOTE:

For a single algebraic term: $x^2 y^3$ the power of x is 2, y is 3. The power of the complete term $x^2 y^3$ is 5 or the sum of the individual powers. (Also referred to as the **degree** of the term $x^2 y^3$ is 5).

SUMMARY OF DEFINITIONS FOR ROOTS AND POWERS:

A **factor** written with an **exponent** produces a **factorization**. The **product** of this factorization is a **power** of the original factor:

$$5^3 = 5 \cdot 5 \cdot 5 = 125$$

factor ↑ | factorization | 3rd power of factor (5)
exponent ↗ | | or product

💡 TO RAISE A NUMBER TO A POWER:

Step 1 Write the number as a factor as many times as indicated by the exponent.

Step 2 Find the product of the factorization.

EXAMPLES:

$2^2 = 2 \cdot 2 = 4$

$3^5 = 3 \cdot 3 \cdot 3 \cdot 3 \cdot 3 = 243$

$10^3 = 10 \cdot 10 \cdot 10 = 1000$

$(\frac{1}{2})^3 = \frac{1}{2} \cdot \frac{1}{2} \cdot \frac{1}{2} = \frac{1}{8}$

$(\frac{3}{4})^4 = \frac{3}{4} \cdot \frac{3}{4} \cdot \frac{3}{4} \cdot \frac{3}{4} = \frac{81}{256}$

$\frac{(5)^2}{(8)^3} = \frac{5 \cdot 5}{8 \cdot 8 \cdot 8} = \frac{25}{512}$

$\frac{5}{3^3} = \frac{5}{3 \cdot 3 \cdot 3} = \frac{5}{27}$

CONCEPT APPLICATION:

The area of a circle is $\pi r^2 = A$

The area of a square is $s^2 = A$

Find the area of the following:

(1) 3 inches on a side of a square = _____

(2) $\frac{1}{2}$ inch on a side of a square = _____

(3) 2.5 inches on a side of a square = _____

(4) The area of a circle is equal to $\pi r^2 = A$. Find the area of a circle with a radius of 2 inches. (π is equal to 3.1416)

= _____

(5) What is the area of a circle with a diameter of 3 inches? = _____ (diameter = 2r)

SQUARE ROOTS:

The square root of a number is **one of the 2 equal factors in its factorization.**

EXAMPLE 1:

The square root of 25 is 5, since the only factorization of 25 that has 2 identical roots is (5 · 5). Our definition says that one of these equal factors is called the square root.

(Written: $\sqrt{25}$ or $\sqrt[2]{25} = 5$)

NOTE:

The symbol "$\sqrt[2]{n}$" is used to indicate the square root of the number it contains, (n). "$\sqrt{}$" is called a **square root** sign or **radical** sign. The number inside the radical is called the **radicand**. The small number "2" is called an **index** number. It is used to determine the number (root) of equal factors in the factorization, one of which will be the correct root. If the index number is omitted we agree to find the square root.

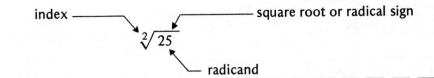

EXAMPLES:

$\sqrt{9} = 3$ $\sqrt{100} = 10$ $\sqrt[3]{27} = 3$

$\sqrt{16} = 4$ $\sqrt[3]{1} = 1$ $\sqrt[4]{16} = 2$

$\sqrt{25} = 5$ $\sqrt[3]{8} = 2$ $\sqrt[4]{81} = 3$

TO FIND A SQUARE ROOT OF A NUMBER to 4 decimal places:

1. Use a square root chart (table).

2. Use a calculator with square root key. (Preferred method — save time and improves accuracy).

CONCEPT APPLICATIONS:

Extract the square root to 4 decimal places:

(1) $\sqrt{256}$ = _____   (6) $\sqrt{100}$ = _____

(2) $\sqrt{.625}$ = _____   (7) $\sqrt{3}$ = _____

(3) $\sqrt{.015625}$ = _____   (8) $\sqrt{2}$ = _____

(4) $\sqrt{27.85}$ = _____   (9) $\sqrt{5}$ = _____

(5) $\sqrt{6400}$ = _____   (10) $\sqrt{7}$ = _____

METRICS

1 = 25.4 mm

UNIT 19
METRICS

INTRODUCTION

The metric measuring system is the most widely used system of measurement throughout the world. It is easy to understand and is convenient to use in mathematical operations. It is not yet widely used in the United States although the number of measurements expressed in metric terms is definitely increasing. Some industries in this country specify dimensions only in metrics on engineering drawings. The major problem for the machinist is that most American made machine tools are calibrated with inch system (U.S. Customary System) readouts. Once the basic conversion factors are learned, both systems can be handled and manipulated with equal ease.

OBJECTIVES:

After completing this unit the student will be able to:

- Convert inch measurements to millimeters
- Convert millimeter measurements to inch measurements
- Solve problems using the metric system of measurement

THE METRIC SYSTEM

The basic unit of length is the meter which is equal to 39.37 inches.

The meter is divided into 100 equal parts called the centimeter which is equal to .3937 inches.

The millimeter is equal to .03937 inches.

There are 25.4 millimeters in an inch.

CONVERSION FORMULA

The most common metric conversions required in shop applications are to change inches to millimeters and millimeters to inches. Convert as follows:

Millimeters = inches × 25.4

Inches = millimeters × .03937

OBSERVE:

The factor 25.4 is the inverse of .03937. That is $\frac{1}{25.4}$ = .03937. You may also see the conversion formula expressed as: mm = in/.03937 and in. = mm/25.4. It is important to remember the relative size of the answers, the millimeter equivalent is larger (by 25.4 times) than the inch equivalent.

INCH/METRIC EQUIVALENT VALUES

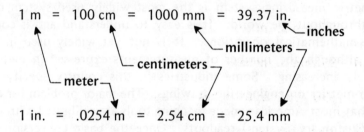

CONCEPT APPLICATIONS:

Convert:

(1) 36 inches to mm

(2) 12 inches to mm

(3) 78.5 inches to mm

(4) 78.5 inches to meters (m)

(5) 1.001 inch to mm

(6) 2 feet, $4\frac{1}{4}$ in. to mm

(7) $3\frac{15}{16}$ in. to mm

(8) 10 mm to inches

(9) 56 mm to inches

(10) 35 mm to inches

(11) 330.708 mm to inches

(12) 10.16 mm to inches

(13) .127 mm to inches

(14) 32.131 mm to inches

(15) 9.525 mm to inches

CONVERTING DIMENSIONS WITHOUT TOLERANCES

RULE:

If the first digit of converted metric value is less than the first digit of the customary value, give the dimension in millimeters to one additional significant digit.

EXAMPLE: Convert 5.4 inches (2 significant digits) to mm.

5.4 in. x 25.4 mm/in. = 137.16 mm, round off to 137 mm (3 significant digits).

RULE:

If the first digit of the converted metric value is greater than the first digit in the customary value, give the dimension in millimeters to the same number of significant digits as the customary value.

EXAMPLE: Convert 2.3 inches (2 significant digits) to millimeters.

2.3 in. x 25.4 mm/in. = 58.42 mm, round off to 58 mm (2 significant digits).

CONCEPT APPLICATIONS:

Convert the following untoleranced dimensions from **inches** to **millimeters**.

(1) .150 = _____   (6) 7.000 = _____

(2) .640 = _____   (7) 5.500 = _____

(3) .863 = _____   (8) 2.875 = _____

(4) 2.480 = _____   (9) 4.515 = _____

(5) 1.002 = _____   (10) 1.750 = _____

ROUNDING TOLERANCED DIMENSIONS

METHOD "A" — NORMALLY PRECISE

Under this method, conversions remain statistically accurate although the limits after conversion may vary by up to 2% of the tolerance. Therefore, using this method, the converted values must be the basis of inspection. The process is:

Step 1 Calculate the maximum and minimum limits in inches.

Step 2 Convert the maximum and minimum limits to millimeter equivalents using the conversion factor, 1 in. = 25.4 mm.

Step 3 Round results to nearest rounded value as indicated in the table below, according to the original tolerance in inches.

Table of Rounding Tolerances
(Inches to Millimeters)

Original Tolerance, in inches		Fineness of Rounding mm
At Least	Less Than	
0.000 01	0.000 1	0.000 01
0.000 1	0.001	0.000 1
0.001	0.01	0.001
0.01	0.1	0.01
0.1	1	0.1

EXAMPLE: METHOD "A"

Convert 1.950 ± 0.016 inches to mm.

The limits are: 1.966 in.
1.934 in.

Convert to mm: 1.966 x 25.4 = 49.9364
1.934 x 25.4 = 49.1236

The tolerance is 0.032 inches (1.966 − 1.934) and lies between 0.01 and 0.1 inch (see table). Therefore, round these values to the nearest 0.01 mm.

The rounded values are 49.94. Note the difference between limits is 0.82.
49.12

The equivalent metric unit and its tolerance is ½ the difference between limits tolerance) and add the tolerance to the lower limit to find the basic unit. Or 49.53 ±0.41.

117

CONCEPT APPLICATIONS:

Convert the following inch dimensions to mm using Method "A".

(1) .375 ± .010

(2) .125 ± .005

(3) .0625 ± .003

(4) .1875 $^{+.0003}_{-.0000}$

(5) .3125 ± .0001

(6) .4375 ± .0015

(7) .500 $^{+.0000}_{-.0001}$

(8) .5625 ± .0025

(9) .625 ± .004

(10) .675 $^{+.001}_{-.000}$

ROUNDING TOLERANCED DIMENSIONS

METHOD "B" — EXTRA PRECISE

This method must be used when tolerances are critical, as for mating parts, or when original inspection equipment is to be used. The process is:

Step 1 Calculate the maximum and minimum limits in inches.

Step 2 Convert the maximum and minimum limits to millimeter equivalents using 1 in. = 25.4 mm.

Step 3 Round each limit to the interior of the tolerance, that is, to the next lower value of the upper limit and the next upper value of the lower limit. Consult the table for the proper number of decimal places.

EXAMPLE: METHOD "B"

Convert the following inch dimension: 1.950 ± 0.016

The limits are: 1.966 in.
1.934 in.

The tolerance is 0.032 inch (see table).

Convert to mm: 1.966 x 25.4 = 49.9364
1.934 x 25.4 = 49.1236

Round to the nearest interior 0.01 mm:

49.93
49.13   or,   49.53 ± 0.40 mm

CONCEPT APPLICATIONS:

Convert the following inch dimensions to mm using Method "B".

(1) 1.094 ± .002

(2) 1.040 +.000 / −.003

(3) .0979 +.0085 / −.0000

(4) .0797 +.006 / −.000

(5) .863 +.003 / −.002

(6) .048 +.002 / −.000

(7) .055 +.004 / −.000

(8) 1.726 ± .003

(9) 1.030 ± .010

(10) 1.2502 ± .0005

CONCEPT APPLICATIONS:

Convert the following inch dimensions to mm using method "A" and then Method "B".

(1) .750 ± .006

(2) .8125 ± .0007

(3) 1.875 ± .001

(4) 5.9375 ± .0004

(5) 1.250 $^{+.0003}_{-.0000}$

(6) 3.0312 ± .005

(7) 1.1093 $^{+.0000}_{-.0001}$

(8) 1.2031 ± .004

(9) 2.718 $^{+.0001}_{-.0000}$

(10) 1.000 ± .0004

(11) .718 $^{+.000}_{-.005}$

(12) .281 ± .002

(13) 1.2054 ± .0005

(14) .749 $^{+.000}_{-.004}$

(15) .8748 $^{+.0000}_{-.0005}$

(16) 9.375 ± .010

(17) .500 $^{+.030}_{-.000}$

(18) .252 ± .001

ROUNDING TOLERANCED DIMENSIONS WHEN CONVERTING FROM METRIC TO INCH UNITS

The most frequent metric conversion in the shop is to convert from metric units to inches. This method can be used to maintain statistically accurate measurements when making the conversion from millimeters to inches. The process is:

Step 1 Calculate the maximum and minimum limits in millimeters.

Step 2 Convert these values exactly into inches by using the conversion formula (inches = .03937 x millimeters).

Step 3 Round the results to the nearest value indicated in the table below.

ORIGINAL TOLERANCE IN mm		ROUND TO (n) places
At Least	Less than	(n) Places in inches
.01 mm	.1 mm	(4) places
.1 mm	1.0 mm	(3) places

CONCEPT APPLICATIONS:

Convert to inches:

1. .19 ± .02 mm

2. .95 +.03 mm / −.01 mm

3. .5 +.1 mm / −.2 mm

4. .6 +.0 mm / −.2 mm

5. .80 +.06 mm / −.00 mm

6. .51 ± .02 mm

7. 2.5 +.01 mm / −.02 mm

8. 40.1 +.00 mm / −.01 mm

9. 23 ± .3 mm

10. 27.1 ± .05 mm

FRACTION-DECIMAL-MILLIMETER EQUIVALENT CHART

Fraction				Decimal	Millimeters	Fraction				Decimal	Millimeters
8ths	16ths	32nds	64ths			8ths	16ths	32nds	64ths		
			1	0.015625	0.396875				33	0.515625	13.096875
		1		.031250	0.793750			17		.531250	13.493750
			3	.046875	1.190625				35	.546875	13.890625
	1			.062500	1.587500		9			.562500	14.287500
			5	.078125	1.984375				37	.578125	14.684375
		3		0.093750	2.381250			19		0.593750	15.081250
			7	.109375	2.778125				39	.609375	15.478125
1				.125000	3.175000	5				.625000	15.875000
			9	.140625	3.571875				41	.640625	16.271875
		5		.156250	3.968750			21		.656250	16.668750
			11	0.171875	4.365625				43	0.671875	17.065625
	3			.187500	4.762500		11			.687500	17.462500
			13	.203125	5.159375				45	.703125	17.859375
		7		.218750	5.556250			23		.718750	18.256250
			15	.234375	5.953125				47	.734375	18.653125
2				0.250000	6.350000	6				0.750000	19.050000
			17	.265625	6.746875				49	.765625	19.446875
		9		.281250	7.143750			25		.781250	19.843750
			19	.296875	7.540625				51	.796875	20.240625
	5			.312500	7.937500		13			.812500	20.637500
			21	0.328125	8.334375				53	0.828125	21.034375
		11		.343750	8.731250			27		.843750	21.431250
			23	.359375	9.128125				55	.859375	21.828125
3				.375000	9.525000	7				.875000	22.225000
			25	.390625	9.921875				57	.890625	22.621875
		13		0.406250	10.318750			29		0.906250	23.018750
			27	.421875	10.715625				59	.921875	23.415625
	7			.437500	11.112500		15			.937500	23.812500
			29	.453125	11.509375				61	.953125	24.209375
		15		.468750	11.906250			31		.968750	24.606250
			31	.484375	12.303125				63	.984375	25.003125
4				.500000	12.700000	8	16	32	64	1.000000	25.400000

NOTE: Table is exact; all figures beyond the six places given are zeros.
BASIS: 1 inch = 25.4 millimeters provides exact six place values.

CONVERSION TABLE — Millimeters to Inches

mm	in	mm	in	mm	in	mm	in
1	0.0394	51	2.0079	101	3.9764	501	19.7244
2	0.0787	52	2.0472	102	4.0157	502	19.7638
3	0.1181	53	2.0866	103	4.0551	503	19.8031
4	0.1574	54	2.1260	104	4.0945	504	19.8425
5	0.1969	55	2.1645	105	4.1339	505	19.8819
6	0.2362	56	2.2047	106	4.1732	506	19.9213
7	0.2756	57	2.2441	107	4.2126	507	19.9606
8	0.3150	58	2.2835	108	4.2520	508	20.0000
9	0.3543	59	2.3228	109	4.2913	509	20.0394
10	0.3937	60	2.3622	200	7.8740	600	23.6220
11	0.4331	61	2.4016	201	7.9134	601	23.6614
12	0.4724	62	2.4409	202	7.9528	602	23.7008
13	0.5118	63	2.4803	203	7.9921	603	23.7402
14	0.5512	64	2.5197	204	8.0315	604	23.7795
15	0.5906	65	2.5591	205	8.0709	605	23.8189
16	0.6299	66	2.5984	206	8.1102	606	23.8583
17	0.6693	67	2.6378	207	8.1496	607	23.8976
18	0.7087	68	2.6772	208	8.1890	608	23.9370
19	0.7480	69	2.7165	209	8.2283	609	23.9764
20	0.7874	70	2.7559	300	11.8110	700	27.5591
21	0.8268	71	2.7953	301	11.8504	701	27.5984
22	0.8661	72	2.8346	302	11.8898	702	27.6378
23	0.9055	73	2.8740	303	11.9291	703	27.6772
24	0.9449	74	2.9134	304	11.9685	704	27.7165
25	0.9843	75	2.9528	305	12.0079	705	27.7559
26	1.0236	76	2.9921	306	12.0472	706	27.7953
27	1.0630	77	3.0315	307	12.0866	707	27.8346
28	1.1024	78	3.0709	308	12.1260	708	27.8740
29	1.1417	79	3.1102	309	12.1654	709	27.9134
30	1.1811	80	3.1496	400	15.7480	800	31.4961
31	1.2205	81	3.1890	401	15.7874	801	31.5354
32	1.2598	82	3.2283	402	15.8268	802	31.5748
33	1.2992	83	3.2677	403	15.8661	803	31.6142
34	1.3386	84	3.3071	404	15.9055	804	31.6535
35	1.3780	85	3.3465	405	15.9449	805	31.6929
36	1.4173	86	3.3858	406	15.9843	806	31.7323
37	1.4567	87	3.4252	407	16.0236	807	31.7717
38	1.4961	88	3.4646	408	16.0630	808	31.8110
39	1.5354	89	3.5039	409	16.1024	809	31.8504
40	1.5748	90	3.5433	500	19.6850	900	35.4331
41	1.6142	91	3.5827			901	35.4724
42	1.6535	92	3.6220			902	35.5118
43	1.6929	93	3.6614			903	35.5512
44	1.7323	94	3.7008			904	35.5906
45	1.7717	95	3.7402			905	35.6299
46	1.8110	96	3.7795			906	35.6693
47	1.8504	97	3.8189			907	35.7087
48	1.8898	98	3.8583			908	35.7480
49	1.9291	99	3.8976			909	35.7874
50	1.9685	100	3.9370			1000	39.3701

(Based on conversion factor 1 in. = 25.4 mm)

CONVERSION TABLE — Inches in Decimals to Millimeters

in	mm	in	mm	in	mm
0.001	0.0254	0.01	0.254	0.1	2.54
0.002	0.0508	0.02	0.508	0.2	5.08
0.003	0.0762	0.03	0.762	0.3	7.62
0.004	0.1016	0.04	1.016	0.4	10.16
0.005	0.1270	0.05	1.270	0.5	12.70
0.006	0.1524	0.06	1.524	0.6	15.24
0.007	0.1778	0.07	1.778	0.7	17.78
0.008	0.2032	0.08	2.032	0.8	20.32
0.009	0.2286	0.09	2.286	0.9	22.86

in	mm	in	mm	in	mm	in	mm
1	25.4	26	660.4	51	1295.4	76	1930.4
2	50.8	27	685.8	52	1320.8	77	1955.8
3	76.2	28	711.2	53	1346.2	78	1981.1
4	101.6	29	736.6	54	1371.6	79	2006.8
5	127.0	30	762.0	55	1397.0	80	2032.0
6	152.4	31	787.4	56	1422.4	81	2057.4
7	177.8	32	812.8	57	1447.8	82	2082.8
8	203.2	33	838.2	58	1473.2	83	2108.2
9	228.6	34	863.6	59	1498.6	84	2133.6
10	254.0	35	889.0	60	1524.0	85	2159.0
11	279.4	36	914.4	61	1549.4	86	2184.4
12	304.8	37	939.8	62	1574.8	87	2209.8
13	330.2	38	956.2	63	1600.2	88	2235.2
14	355.6	39	990.6	64	1625.6	89	2260.6
15	381.0	40	1016.0	65	1651.0	90	2286.0
16	406.4	41	1041.4	66	1676.4	91	2311.4
17	431.8	42	1066.8	67	1701.8	92	2336.8
18	457.2	43	1092.2	68	1727.2	93	2362.2
19	482.6	44	1117.6	69	1752.6	94	2387.6
20	508.0	45	1143.0	70	1778.0	95	2413.0
21	533.4	46	1168.4	71	1803.4	96	2438.4
22	558.8	47	1193.8	72	1828.8	97	2463.8
23	584.2	48	1219.2	73	1854.2	98	2489.2
24	609.6	49	1244.6	74	1879.6	99	2514.6
25	635.0	50	1270.0	75	1905.0	100	2540.0

EXAMPLE: 1.25″ = ___?___ mm

       1. = 25.4
       .2 = 5.08
      .05 = 1.27
           31.75 ans

ALGEBRA

$$4ab=y$$

UNIT 20

ALGEBRA - ADDITION AND SUBTRACTION

INTRODUCTION

Algebra is a generalized arithmetic form that uses symbols and letters to represent numbers or variables. A simple expression in algebraic form represents many different answers depending upon what value is assigned to each variable in the expression. A set of simple rules will allow you to add and subtract terms expressed in algebra. Just keep in mind that one apple + one orange = one apple + one orange, and nothing else. However, if you have one apple + one orange in a bag and one apple + one orange in another bag, the total = two apples + two oranges + two bags. Keep this basic concept in mind: you can compare equal objects with equal objects but you cannot compare unequal or unlike objects.

OBJECTIVE:

After completing this unit the student will be able to:

- Define the basic terms of an algebraic expression
- Add algebraic terms
- Subtract algebraic terms

DEFINITIONS:

ALGEBRAIC EXPRESSION

Contains combinations of numbers, signs of operation, and letters (variables):

EXAMPLES:

$2abc \qquad x + y \qquad a^2 + 4 = 8$

NOTE:

- **Numbers** have a definite (known) value.

- **Letters(variables)** are used to replace numbers whose value is not given or is unknown.

- A single letter "a" means "1 • a" but the "1" is not necessarily written.

- **Signs of Operation** "+" or "−" separate **Terms**. (a + b + c − d) has 4 terms.

ALGEBRAIC TERMS

An algebraic term is a single expression containing the following parts.

CONSIDER:

Number → $5a^3$ means → (5 • a • a • a)

with Exponent and Letter labeled.

Number — called **coefficient** — may not be written in all terms if the value is 1. ("a" means "1 • a")

Letters(s) — called **variable(s)** — may contain any number of letters: (5abcd)

Exponent — tells how many times the **variable(s)** is to be used as a factor: $5a^2 b^2 c^3 d = 5 • a • a • b • b • c • c • c • d$

LIKE TERMS

Two or more terms containing the **same letters** with the **identical exponents**.

EXAMPLES:

$5ab^2$ and $7ab^2$

NOTE:

The numbers are not required to be the same, but the letters "a" and "b" must both exist, and the exponent "2" must be attached to both "b" factors. LIKE TERMS may be added or subtracted.

UNLIKE TERMS

Two or more terms that differ because they have different letters (bases) and/or different exponents.

EXAMPLES:

$5a$ and $6b$ have different bases
$5a$ and $6a^2$ have different exponents

NOTE:

UNLIKE terms cannot be combined by adding or subtracting. (However, it is generally possible to combine them by multiplication and division — to be developed later).

TO ADD ALGEBRAIC TERMS

Step 1 Combine the numbers by addition using the rules for signed numbers.

Step 2 Write the common base and exponents.

EXAMPLES:

a. $3a^4 + 4a^4 = 7a^4$

b. $a + 2a + 3a = 6a$ $(a = 1 \cdot a)$

c. $2a + 4b + 3c$
       $5a - 2b + 4c$
       ―――――――――
       $7a + 2b + 7c$

d.   add: $2x^2 + x - 3y + y^2$ and $2x - y + 2y^3$

$2x^2$	$+\ x$	$-\ 3y$	$+\ y^2$	
	$2x$	$-\ y$		$+\ 2y^3$
$2x^2$	$+\ 3x$	$-\ 4y$	$+\ y^2$	$+\ 2y^3$

NOTE:

Terms are arranged in columns of LIKE terms only. UNLIKE terms are alone in a column.

CONCEPT APPLICATIONS:

Add:

(1) $a + b$
$\ \ a + b$

(2) $m - n$
$\ \ m + n$

(3) $x - y$
$\ \ x - y$

(4) $7b + 3c$
$ -2b - 7c$

(5) $x + 2b$
$\ \ x + b$

(6) $4c + 5d$
$ -c + 8d$

(7) $-3m + 12n$
$\ 11m - 15n$

(8) $9b + 8c$
$ -3b - 6c$

(9) $3x + y$
$\ \ x - y$

(10) $10x^2 - 9y^2$
$\ \ 4x^2 - 6y^2$

(11) $5a^2 + 3b^2$
$\ 2a^2 - 3b^2$

(12) $3x^2 - y^2$
$ -2x^2 + 5y^2$

(13) $5m + 2n - 3r$
$\ \ m - n + 4r$

(14) $-15s + 7t - u$
$\ \ \ 8s - 3t + 2u$

(15) $3a - 5b + 4c$
$\ 3a + 8b - 6c$

(16) $8m + n + p$
$\ 3m - 7n - 2p$

TO SUBTRACT ALGEBRAIC TERMS

Step 1 Combine the numbers by subtraction using the subtraction rule for signed numbers.

Step 2 Write the common base and exponents.

EXAMPLES:

a. $2a - a = a$

b. $12a^3 - 6a^3 = 6a^3$

c. $15ab^2c - 11ab^2c = 4ab^2c$

d.
$$\begin{array}{r} 20x^2 + 14y^2 - 8z^2 \\ \ominus \quad\;\; \oplus \quad\;\; \ominus \\ 14x^2 - 8y^2 + 4z^2 \\ \hline 6x^2 + 22y^2 - 12z^2 \end{array}$$

NOTE:

Recall subtraction rule — change the bottom signs then add.

Subtract $2a + b - 3c$ from $3a + 4b - 2c$

$$\begin{array}{r} 3a + 4b - 2c \\ \ominus \quad \ominus \quad\;\; \oplus \\ 2a + b - 3c \\ \hline a + 3b + c \end{array}$$

CONCEPT APPLICATIONS

Subtract:

(1) $3a$
     $2a$

(2) $5x$
     $4x$

(3) $6m$
     $-m$

(4) $-3b$
     $+2b$

(5) $-4c$
     $-7c$

(6) $2x^2$
     $-x^2$

(7) $-mn$
     mn

(8) $-12t$
     $-10t$

(9) $7abc$
     $9abc$

(10) $3x^2y^2$
     x^2y^2

(11) $2a + 3b$
     $a + b$

(12) $x + 7y$
     $x + 4y$

(13) $4b - 3c$
     $3b - 2c$

(14) $7c + 5d$
     $3c + 5d$

(15) $6a - 5mn$
     $4a - 2mn$

(16) $-7x - 9u$
     $-3x - 7u$

(17) $14 + ab$
     $8 + ab$

(18) $2b + 7c$
     $2b + 5c$

(19) $10 - 6d$
     $10 - 4d$

(20) $3p - 6$
     $p - 7$

(21) a − b
      a + b

(22) 2x − y
      −x − y

(23) 3m + 8n
      2m − 3n

(24) m − 5n
      m + 4n

(25) r + s
      r − s

(26) 5y − 3z
      3y + 2z

(27)   3b − 7c
      −7b + 3c

(28) 2x + 5y
      9x − 4y

(29) 6a − 8b
      6a + 8b

(30) 4y − 9
      9y + 2

(31) 3a + 4b − 2c
       a + 3b − c

(32) 5m + 3n + 4
      3m + 2n − 6

(33) 7x − 5y − 4
      8x − 6y + 5

(34) 8a − 4b + 7c
      2a + 3b + 8c

(35) 7c + d − f
      3c − 7d − 2f

(36)   2u + 4v − 7
      −3u + 7v + 3

(37) 3x − 5x + 4
      2x + 6x − 4

(38)   2 + 4m − 7m^2
      −3 + 7m − 9m^2

(39) a + ab + b^2
      a − ab + b^2

(40) 5a^2b^2 + 3ab − 1
      4a^2b^2 + 4ab + 1

UNIT 21

ALGEBRA - SYMBOLS AND OPERATIONS

INTRODUCTION

Algebra uses symbols, such as parentheses and brackets, that indicate the order in which operations are to be done. Without such guidelines the machinist sometimes would not get the right answer when using a complicated formula. There are hundred of different algebraic formulas applicable to machining. Each of these can be manipulated and solved with ease after you learn to identify the symbols and how to operate on them, step by step, to reach a single solution.

OBJECTIVES:

After completing this unit the student will be able to:

- Simplify expressions containing symbols of inclusion, (   ), [   ], {   }.
- Simplify expressions by applying the order of operations rule.

DEFINITIONS:

SYMBOLS OF INCLUSION

A symbol such as parentheses, brackets or braces, (   ), [   ], {   } is used to group a set of algebraic factors or terms. These are also called grouping symbols.

EXAMPLES:

a. $(2 \cdot 2 \cdot 2)$ — set of (3) factors

b. $(47 + \frac{1}{8})$ — set of (2) terms

> NOTE:
>
> If a **SYMBOL OF INCLUSION** is found inside one or more others, simplify the inner one first. Repeat the process if necessary.

EXAMPLE:

$\{20 + [50 - (25 \div 5)] + 1\} =$ remove ( )

$\{20 + [50 - 5] + 1\} =$ then [ ]

$\{20 + 45 + 1\} =$ then { }

$\{66\} = 66$

PARENTHESES

( ) are used as a basic grouping symbol, but they also serve to show multiplication of terms when two parentheses are written together without an operation sign between them: (5)(7) means: 5 times 7.

BRACKETS and BRACES

[ ] and { } are also used as basic grouping symbols. However, when one term, quantity, etc. is found within another, the innermost symbol is generally the ( ) followed by [ ], then { }.

Solve the following example:

$\{2 [20 - (6 + 8) + 4]^2\} = \underline{\;?\;}$

The answer is 200.

TO REMOVE A SYMBOL OF INCLUSION:

1. If the symbol is preceded by a **plus** or positive sign, remove the symbol and make no further changes:

   $+ (7 + 8) = 7 + 8$ since $(15) = 15$

2. If the symbol is preceded by a **minus** or negative sign, remove the symbol and change the sign of each term inside:

   $- (10 + 6) = -10 - 6 = -(16) = -16$

REMEMBER:

$(-a) = -(a) = -a$ and $(-5) = -(5) = -5$

EXAMPLE 1:

$(2a - 3b) + (a - b)$

$2a - 3b + a - b = 3a - 4b$

Remove the ( ) preceded by a positive sign and collect like terms.

EXAMPLE 2:

$4a - (3c - 2d) + a^2$

$4a - 3c + 2d + a^2$

Remove the ( ) preceded by a negative sign, change the signs of the inner terms and collect like terms.

EXAMPLE 3:

$2ab - 3a - (a + 2) - (a + b) =$

$2ab - 3a - a - 2 - a - b =$

$2ab - 5a - b - 2$

ORDER OF OPERATIONS LAW

> To simplify an expression the following steps are taken:
>
> *Step 1* Remove symbols of inclusion starting with the innermost and complete the operations on terms they contain.
>
> *Step 2* Simplify indicated powers and roots where possible.
>
> *Step 3* Perform the indicated multiplication and division in order from left to right.
>
> *Step 4* Perform the addition and subtraction in order from left to right.

EXAMPLE 1:

$2 \cdot [3 + (4 - 2)] - 5 =$	
$2 \cdot [3 + 4 - 2] - 5 =$	remove ( )
$2 \cdot [7 - 2] - 5 =$	combine like terms
$2 \cdot [5] - 5 =$	combine like terms
$2 \cdot 5 - 5 =$	remove [ ]
$10 - 5 = 5$	combine like terms

NOTE:

Certain steps may be done mentally, but use caution.

EXAMPLE 2:

$$x + [y - (x + y)] =$$
$$x + [y - x - y] =$$
$$x + [\ -x\ ] =$$
$$x - x = 0$$

NOTE:

A number **plus** its opposite = 0 $y + (-y) = 0$ and
$n + (-n) = 0$ and
$6 + (-6) = 0$

CONCEPT APPLICATIONS:

Simplify using the Order of Operations Law:

(1) $(2a + b) + (a + 2b)$

(2) $(3m - n) - (m + 2n)$

(3) $(x + 5y) + (3x - 7y)$

(4) $(a + b - c) - (a - b - c)$

(5) $(m - n + p) - (m - n + p)$

(6) $(4x - 2y + z) - (3x - 2y - z)$

(7) $x + [y + (x - y)]$

(8) $x + [y - (x + y)]$

(9) $x - [x + (y - z)]$

(10) $x - [x - (x + y)]$

(11) $a + [a - (a - 1) + 1]$

(12) $a - [a + (a - 1) - 1]$

(13) $m - [m - (m - 1) - 1]$

(14) $3c - [1 - (1 - c) + c]$

(15) $5 - [4 - (8 + d)]$

(16) $4d - [2b + (b - 2)]$

(17) $3b - [(b - 4) - 1]$

(18) $2x + [7 - (3 - x)]$

(19) $m - [(m - n) - (m + n)]$

(20) $6x - [(5x - y) + (2x + y)]$

UNIT 22

ALGEBRAIC MULTIPLICATION

INTRODUCTION

Machinists' handbooks and reference books contain many general purpose formulas. For the skilled machinist, the practical application for algebraic multiplication includes applying formulas to specific operations that are part of everyday work. Examples involve calculating the surface speed for a cutting operation, the shape of a gear tooth or the layout of a bolt circle.

OBJECTIVE:

After completing this unit the student will be able to:

- Multiply two or more algebraic terms
- Multiply a term by a quantity
- Multiply a quantity by a quantity

DEFINITIONS:

✋ TERM
> a single algebraic expression in the form of 2, x, 2x, $-2x$, $2x^2$, $-2x^2$, 5abc, etc.

✋ QUANTITY
> an expression containing two or more TERMS: $(2 + 5)$, $(x + y)$, $(2x^2 + 3y^3)$, $(1 + a - b)$, etc.

TO MULTIPLY TWO OR MORE TERMS:

Step 1 Multiply the coefficient (number) of the first term by the coefficient of the second term.

Step 2 Multiply the variable (letter) of the first term by the variable of the second term.

EXAMPLE 1:

$$2x^2 \cdot 3x^3 = 2 \cdot 3 \; (x \cdot x)(x \cdot x \cdot x)$$
$$= 6 \; x \cdot x \cdot x \cdot x \cdot x$$
$$= 6 \; x^5$$

EXAMPLE 2:

$$-2ab \cdot 4ab = -2 \cdot 4ab \cdot ab = -2 \cdot 4 \cdot ab \cdot ab$$
$$= -8a^2b^2$$

EXAMPLE 3:

$$2x^2y^3 \cdot 3x^3y^4 = 2 \cdot 3 \cdot x^2 \cdot y^3 \cdot x^3 \cdot y^4$$

(rearrange factors)

$$= 6 \, (x^2 x^3)(y^3 y^4)$$
$$= 6 \, x^5 y^7$$

RULE:

$$x^a \cdot x^b = x^{a+b} \quad \text{and} \quad \frac{x^a}{x^b} = x^{a-b}$$

NOTE:

$\quad x^2 \cdot x^3 = x^5 \quad$ It is **not** x^6, ($\neq x^6$)

OBSERVE:

Substitute $x = 2$ in the following:

$$x^2 \cdot x^3 = x^5$$
$$(2 \cdot 2)(2 \cdot 2 \cdot 2) = (2 \cdot 2 \cdot 2 \cdot 2 \cdot 2)$$
$$4 \cdot 8 = 32$$
$$32 = 32$$

TO MULTIPLY A TERM BY A QUANTITY

Step 1 Multiply the term by each term contained in the quantity.

Step 2 Simplify the product when possible. (Add LIKE Terms)

EXAMPLE 1:

$$4(5 + 6) = 4(5 + 6) \quad \text{Compare:} \quad a(b + c) = a \cdot b + a \cdot c$$
$$= 4 \cdot 5 + 4 \cdot 6 \qquad\qquad\qquad = ab + ac$$
$$= 20 + 24$$
$$= 44$$

EXAMPLE 2:

$$2x(4x - 3y + 2) = 8x^2 - 6xy + 4x \quad \leftarrow \text{Horizontal Form}$$

$$\begin{array}{r} 4x - 3y + 2 \\ 2x \\ \hline 8x^2 - 6xy + 4x \end{array} \quad \downarrow \text{Vertical Form}$$

MULTIPLY A QUANTITY BY A QUANTITY:

Step 1 Multiply each term of the first quantity by each term of the second quantity.

Step 2 Simplify the product when possible.

EXAMPLE 1:

$$(x + 2)(x + 5) = (x + 2)(x + 5) = \underline{\quad ? \quad}$$

Step 1	$x(x + 5) = x^2 + 5x$		or	$x + 2$
Step 2	$2(x + 5) = 2x + 10$			$\underline{x + 5}$
(add LIKE terms)	$= x^2 + 7x + 10$			$x^2 + 2x$
				$\underline{ 5x + 10}$
				$x^2 + 7x + 10$

EXAMPLE 2:

$$(a + b)^2 = (a + b)(a + b) = \begin{array}{r} a^2 + ab \\ ab + b^2 \\ \hline a^2 + 2ab + b^2 \end{array}$$

or
$$\begin{array}{r} a + b \\ a + b \\ \hline a^2 + ab \\ ab + b^2 \\ \hline a^2 + 2ab + b^2 \end{array}$$

EXAMPLE 3:

$$(a + b)(a - b) = \begin{array}{r} a^2 - ab \\ + ab - b^2 \\ \hline a^2 - b^2 \end{array} \qquad \text{or} \quad \begin{array}{r} a + b \\ a - b \\ \hline a^2 + ab \\ - ab - b^2 \\ \hline a^2 - b^2 \end{array}$$

EXAMPLE 4:

$$5b^2 - [3b(b - 8)] =$$
$$5b^2 - [3b^2 - 24b] =$$
$$5b^2 - 3b^2 + 24b =$$
$$2b^2 + 24b$$

CONCEPT APPLICATIONS:

Set 1

(1) $3(2a) = $ _____

(2) $(-2)(+3a) = $ _____

(3) $3a(-4c) = $ _____

(4) $(-x)(mn) = $ _____

(5) $(5x)(-7x) = $ _____

(6) $3(2a + 1) = $ _____

(7) $-3(b - 4) = $ _____

(8) $-4(c + 7) = $ _____

(9) $2x(x - 4) = $ _____

(10) $-x(-x + 1) = $ _____

Set 2

(1) $(a + b)(a - b) = $ _____

(2) $(m + 2)(m - 2) = $ _____

(3) $(x - 5)(x + 5) = $ _____

(4) $(y + 4)(y - 4) = $ _____

(5) $(n + 3)(n - 3) = $ _____

(6) $(b + 2c)(b - 2c) = $ _____

(7) $(r + 6s)(r - 6s) = $ _____

(8) $(u - 4v)(u + 4v) = $ _____

(9) $5c - 2(c - 1) = $ _____

(10) $x^2 + x(x + 1) = $ _____

(11) $x^2 - x(x - 1) = $ _____

(12) $5b^2 - 3b(b - 8) = $ _____

(13) $(2m + 3n)(2m - 3n) = $ _____

(14) $(3a - 7b)(3a + 7b) = $ _____

(15) $(5c + 6d)(5c - 6d) = $ _____

(16) $(8d - 1)(8d + 1) = $ _____

(17) $\left(\frac{1}{2}x - y\right)\left(\frac{1}{2}x + y\right) = $ _____

(18) $\left(3c + \frac{1}{2}d\right)\left(3c - \frac{1}{2}d\right) = $ _____

(19) $\left(mn + \frac{3}{4}\right)\left(mn - \frac{3}{4}\right) = $ _____

(20) $\left(\frac{1}{2}x - \frac{1}{3}y\right)\left(\frac{1}{2}x + \frac{1}{3}y\right) = $ _____

(21) $(2b + 1.5)(2b - 1.5) = $ _____

(22) $(2.5 - 3m)(2.5 + 3m) = $ _____

Set 3 Write the square of each of the following:

(1) $(a + b)^2 =$

(2) $(c + d)^2 =$

(3) $(m + n)^2 =$

(4) $(x + y)^2 =$

(5) $(r - s)^2 =$

(6) $(a - b)^2 =$

(7) $(c + 4)^2 =$

(8) $(d + 3)^2 =$

(9) $(m - 5)^2 =$

(10) $(y - 6)^2 =$

(11) $(r - 4)^2 =$

(12) $(b + 5)^2 =$

Set 4

(1) $(4m + 5n)^2 =$

(2) $(3a - 4c)^2 =$

(3) $(5c - 6d)^2 =$

(4) $(7b + 8c)^2 =$

(5) $(2y + 9z)^2 =$

(6) $(3m - 9n)^2 =$

(7) $(8d + f)^2 =$

(8) $(7a - b)^2 =$

(9) $(6b - c)^2 =$

(10) $(r + 5s)^2 =$

(11) $(3a + 2b)^2 =$

(12) $(2n + 3y)^2 =$

UNIT 23

SUBSTITUTING NUMERICAL VALUES

INTRODUCTION

A formula is a general mathematical statement or rule. The formula may contain letters, numbers, signs of operation and an equality symbol. The machinist oftens uses a reference handbook to find the right formula to perform a specific job. Numerical values relating to the specific job are substituted into the appropriate formula to produce the measure or value needed to perform the job.

OBJECTIVE:

After completing this unit the student will be able to:

- Solve equations by substituting numerical value for the variables.
- Evaluate the numerical value of an algebraic expression.

DEFINITIONS

EQUATION
An expression stating two things are equal: $2 + 3 = 5$

FORMULA
A rule written in the form of an equation: $A = s^2$

TO EVALUATE
To determine the numerical value of an algebraic expression.

EXAMPLE 1:

Given: If a = 5, b = 2, find the value of:

$$(2a + b) \div 2b =$$

Substitute: $(2 \cdot 5 + 2) \div 2 \cdot 2 =$

$$12 \div 4 = 3$$

or $(2a + b) \div 2b = 3$ If: a = 5 and b = 2

EXAMPLE 2:

The formula used to determine **taper per inch** is:

$TPI = \dfrac{D - d}{L}$, which relates to this figure:

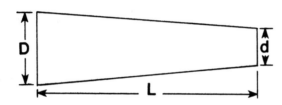

Solution: If D = 1.40 inches
d = 1.20 inches
L = 2   inches, then

$TPI = \dfrac{D - d}{L}$ or

$TPI = \dfrac{1.40 - 1.20}{2} = \dfrac{.20}{2} = .10$ or .1 inch

- -

NOTE:

Both of these examples require the use of the ORDER OF OPERATIONS LAW. Review this process from Unit 21 before attempting the following exercises.

- -

CONCEPT APPLICATIONS:

Set 1

Find the value of the following: Given: a = 6, b = 4, c = 2, x = 1, and y = 5

(1) 5y − 4a =

(2) 4b − 3y =

(3) x + 4y =

(4) b + 7c =

(5) 3c − 3x =

(6) a + 2c =

(7) 3a + x =

(8) 5a + y =

(9) 3c + b =

(10) 3a + 5x =

(11) 2b + 3a =

(12) 4c − a =

(13) 2b − 4c =

(14) 10x − y =

(15) 3b − 2a =

(16) 7a − 9b =

(17) 10c + y =

(18) 9y − 7a =

(19) 15x − 2y =

(20) 6b − 2a =

Set 2

Find the values of the following: Given: a = 2, b = 5, c = 6, x = 3, y = 4 and z = 1

(1) $\dfrac{a + c}{4} =$

(2) $\dfrac{b + x}{2} =$

(3) $\dfrac{a + y}{6} =$

(4) $\dfrac{b - a}{3} =$

(5) $\dfrac{c - a}{2} =$

(6) $\dfrac{a + b + z}{4} =$

(7) $\dfrac{x + y - z}{3} =$

(8) $\dfrac{c - b + x}{2} =$

(9) $\dfrac{b + c - x}{4} =$

(10) $\dfrac{c + y + b}{5} =$

(11) $\dfrac{a + b}{c + z} =$

(12) $\dfrac{c - x}{c + x} =$

(13) $\dfrac{y + a}{y - a} =$

(14) $\dfrac{2c + y}{b + x} =$

(15) $\dfrac{4x - 2y}{a + c} =$

(16) $\dfrac{a + c + z}{c - a} =$

(17) $\dfrac{b + c - y}{x + z} =$

(18) $\dfrac{2a + b + y}{a + x + z} =$

(19) $\dfrac{5x + 2b - y}{3a + z} =$

(20) $\dfrac{4c + 3x + 2z}{9b + x - 3a} =$

Set 3

(1) Find H when E = −6, F = 9, G = −2

$$H = E + F + G$$

(2) Find Y when P = −2, x = −5, A = −1

$$Y = 6 + P + X + A$$

(3) Find P when S = −7, T = −5, W = −3

$$P = S - T + W$$

(4) Find R when A = 3, B = −14, C = 5

$$R = 2A - B - C$$

(5) Find Y when S = 2, P = 7, N = −25

$$Y = 2SP + N - S - P$$

(6) Find X when F = 3, G = −17, H = −5

$$X = 2F + G - H$$

(7) Find X in #6 when G = −7, H = −9 and F = 8

$$X = 2F + G - H =$$

(8) Find P when A = −2, B = −3, C = −6, D = −8

$$P = A - B + C - D$$

Set 4

(1) Find H when E = 4, F = 5, G = 9
$$H = E + F + G$$

(2) Find Y when P = 3, A = 5, X = 2
$$Y = 17 + P + X + A$$

(3) Find Y when R = 5, N = 7
$$Y = \frac{R}{N}$$

(4) Find A when B = 3, C = 4, D = 6
$$A = 5 + BC + CD + BD + 2$$

(5) Find P when S = 3, W = 8, T = 6
$$P = \frac{1}{ST} + 5SW - \frac{TW}{S}$$

(6) Find Y when A = 2, P = 6, D = 3, H = 4
$$Y = 10 - \frac{PH}{D} + PDA + \frac{AP}{DH}$$

(7) Find A when B = 5, C = 3, D = 4, E = 2, F = 1, G = 7, H = 12
$$A = (B - C)D - \frac{E}{F} + G - H$$

(8) Find A in formula #7, when B = 5, C = −3, D = 4, E = 2, F = 1, G = −7, H = −12

151

Set 5

(1) Find X when A = 3, H = 5, N = 4
$$X = A[N - 2(H - A)]$$

(2) Find W when L = 5, X = 6
$$W = L[3 + 4(X + 1)^2]$$

(3) Find X when P = 3, Y = 4
$$X = P\left[(Y - 1)\frac{(P + 3)}{Y}\right]$$

(4) Find M when X = 4, Y = 11
$$M = X[3(X + 1)^2 - (Y - 3)^2]$$

(5) Find X when A = −6, B = 4
$$X = A[B(3 - A)^2 - (B - 2)^2]$$

(6) Find Y when A = 6, B = −2, C = −5
$$Y = BC\{A - B[(ABC + C)(A - B)^2]\}$$

UNIT 24

EQUATIONS - ADDITION AND SUBTRACTION PRINCIPLES

INTRODUCTION

The equation is an expression of equality. An equation has an equal (=) sign placed so that the terms to the left will exactly equal in value the terms to the right of the equal sign. There are rules to follow that ensure the equation remains in balance when you add values, or subtract values, from both sides of the balance (=) sign.

OBJECTIVES:

After completing this unit the student will be able to:

- Solve linear equations using the addition principles of equality
- Solve linear equations using the subtraction principle of equality

PRINCIPLES/RULES:

PRINCIPLE OF EQUALITY:
> To maintain the equality it is necessary to perform the same operation on both sides of the equation using the same number or term of equal value.

ADDITION PRINCIPLE OF EQUALITY:
> If the same number or term is added to each side of the equation, the two sides will remain equal.
>
> $$\text{equals} + \text{equals} \longrightarrow \text{equals}$$
> $$(\ =\ +\ =\ \text{will produce}\ =\ )$$

EXAMPLE 1:

```
        Vertical ↓                    Horizontal →
        10   =   10              10        =   10
       + 5   =  + 5               10 + 5   =   10 + 5
       ---      ---                  15    =   15
        15   =   15
```

EXAMPLE 2:

```
           Vertical ↓                      Horizontal →
     x + 40  =  x + 40          x  +  40     =   x  +  40
         + 10 =     + 10         __ + __ + 10 = __ + __ + __
     ---------   ---------
     x + 50  =  x + 50            __ + __    =  __ + __
```

Solve the equations for a value of x. (x = __?__)

EXAMPLE 3:

```
    x − 20 = 45  ⟶   x − 20  =   45
                     +  20   =  +20     add 20 to each side
                     ---------   -----
                         x    =   65
```

EXAMPLE 4:

```
    x − .5 = 3.5  ⟶   x −  .5  =  3.5
                      +   .5  =  +.5    add .5 to each side
                      ---------   -----
                          x   =  4.0
                          x   =  4
```

💡 SUBTRACTION PRINCIPLE OF EQUALITY:

If the same number or term is subtracted from each side of the equation, the two sides will remain equal.

equals − equals ⟶ equals

(= − = will produce =)

EXAMPLE 1:

```
        Vertical ↓                      Horizontal →
        13   =   13              13            =   13
       − 7   =  − 7               13  −  7     =   13  −  7
       ---      ---                   6        =       6
         6   =    6
```

EXAMPLE 2:

$$x + \frac{3}{8} = x + \frac{3}{8}$$
$$\underline{-\frac{3}{8} = -\frac{3}{8}}$$
$$x = x$$

$$x + \frac{3}{8} = x + \frac{3}{8}$$
$$\underline{ + + (-\frac{3}{8}) = + + }$$
$$\underline{} = \underline{}$$

Solve the equations for a value of x. (x = __?__)

EXAMPLE 3:

$$x + 8 = 18 \longrightarrow \begin{array}{rcr} x + 8 &=& 18 \\ -8 &=& -8 \\ \hline x &=& 10 \end{array}$$

EXAMPLE 4:

$$x - .645 = .354 \longrightarrow \begin{array}{rcr} x - .645 &=& .354 \\ +.645 &=& +.645 \\ \hline x &=& .999 \end{array}$$

Using the horizontal form of addition and subtraction

EXAMPLE 5:

$$x + .34 = .45$$
$$x \boxed{+ .34 - .34} = .45 - .34$$
$$x = .11$$

EXAMPLE 6:

$$x - \frac{13}{16} = \frac{15}{16}$$
$$x \boxed{- \frac{13}{16} + \frac{13}{16}} = \frac{15}{16} + \frac{13}{16}$$
$$x = \frac{28}{16}$$
$$x = \frac{14}{8} = 1\frac{6}{8} = 1\frac{3}{4} \quad \text{(simplified)}$$

this quantity $\boxed{}$ = 0 in both examples

155

CONCEPT APPLICATIONS:

Using the principles of equality for addition and subtraction solve the following:

Addition **Subtraction**

(1) x − 1 = 2; x = _____   (15) n + 4 = 8; n = _____

(2) x − 2 = 7; x = _____   (16) n + 2 = 8; n = _____

(3) m − 5 = 8; m = _____   (17) 35 = x + 50; x = _____

(4) d − 7 = 7; d = _____   (18) 75 = y + 25; y = _____

(5) y − 2 = 5; y = _____   (19) 0 = m + 30; m = _____

(6) n − 3 = 6; n = _____   (20) 40 = n + 50; n = _____

(7) c − 7 = 6; c = _____   (21) x + 13 = 25; x = _____

(8) b − 8 = 8; b = _____   (22) y + 19 = 19; y = _____

(9) b − 8 = 4; b = _____   (23) b + 25 = 15; b = _____

(10) x − 5 = 2; x = _____   (24) c + 15 = 25; c = _____

(11) x − 2 = 5; x = _____   (25) n + 18 = 6; n = _____

(12) y − 2 = 1; y = _____   (26) p + 24 = 32; p = _____

(13) y − 1 = 4; y = _____   (27) 24 = y + 32; y = _____

(14) r − 6 = 3; r = _____

UNIT 25

EQUATIONS - MULTIPLICATION AND DIVISION PRINCIPLES

INTRODUCTION

The balance of equality of an equation can be maintained when multiplying or dividing on both sides. This is similar to the way addition and subtraction operations were handled in Unit 24. These multiplication and division principles are especially useful in solving problems in trigonometry where formulas are used to find angles and measures.

OBJECTIVES:

After completing this unit the student will be able to:

- Solve linear equations using the multiplication and the division principles of equality
- Solve linear equations which require two or more principles of equality

PRINCIPLES/RULES:

💡 MULTIPLICATION PRINCIPLE OF EQUALITY:

If both sides of the equation are multiplied by the same number, both sides will remain equal.

EXAMPLE 1:

If a number (n) is divided by 4, the result (quotient) is 7. Express as an equation and solve for the value of (n) = __?__ .

therefore: $\dfrac{n}{4} = \dfrac{7}{1}$,

Multiply both side by the denominator "4", then **cancel**.

$$\frac{4}{1} \cdot \frac{n}{4} = \frac{4}{1} \cdot \frac{7}{1}$$

$$1n = \frac{28}{1} = 28$$

$$n = 28$$

EXAMPLE 2:

$$\frac{n}{6} = 9$$

$$6 \cdot \frac{n}{6} = 6 \cdot 9$$

$$n = 54$$

DIVISION PRINCIPAL OF EQUALITY:

If both sides of the equation are divided by the same number, both sides will remain equal.

EXAMPLE 1:

$$24x = 48 \quad (\div \text{ by } 24)$$

$$\frac{24x}{24} = \frac{48}{24}^2 \quad \text{cancel/simplify}$$

$$x = 2$$

EXAMPLE 2:

$$4x = 1.6$$

$$\frac{4x}{4} = \frac{1.6}{4}$$

$$x = .4$$

EQUATIONS WHERE TWO OR MORE PRINCIPLES OF EQUALITY APPLY

To solve an equation when two or more "Principles of Equality" are necessary:

Step 1 Apply the Addition or Subtraction rules to eliminate terms.

Step 2 Apply the Multiplication or Division rules to remove the unwanted factors, and unwanted denominators.

EXAMPLE 1:

$$5x - 8 = 12$$ 5 is a factor, 8 is a term,

$$5x - 8 + 8 = 12 + 8$$ remove the factor "8"

$$5x = 20$$ remove the factor "5"

$$\frac{\cancel{5}x}{\cancel{5}} = \frac{\cancel{20}^{4}}{\cancel{5}}$$

$$x = 4$$

EXAMPLE 2:

$$\frac{x}{6} + 8 = 18$$

$$\frac{x}{6} + 8 - 8 = 18 - 8 \quad \text{remove terms}$$

$$\frac{x}{6} = 10$$

$$\frac{\cancel{6}x}{\cancel{6}} = 6 \cdot 10 \quad \text{remove denominators}$$

$$x = 60$$

EXAMPLE 3:

$$\frac{3x}{4} + 3 = 15$$

$$\frac{3x}{4} + 3 - 3 = 15 - 3 \quad \text{remove terms}$$

$$\frac{3x}{4} = 12$$

$$\frac{\cancel{4}}{\cancel{3}} \cdot \frac{\cancel{3}x}{\cancel{4}} = \frac{4}{\cancel{3}} \cdot \cancel{12}^{4} \quad \text{remove factors and denominators}$$

$$x = 4 \cdot 4$$

$$x = 16$$

GOLDEN RULE OF EQUATIONS

Any operation (such as: $+, -, \cdot, \div, (\ )^2$ or $\sqrt{\ }$) may be performed on one side of an equation using any number **if** you perform the same **operation** on the other side with the same number.

TO DETERMINE IF EQUALITY EXISTS:

If $\frac{a}{b} = \frac{c}{d}$ is true, the pair of fractions are equal

Step 1 **Multiply** the first numerator by the second denominator (a x d).

Step 2 **Multiply** the first denominator by the second numerator (b x c).

Step 3 If these two **products** are equal, then the fractions are equal. (That is ad = bc).

This process, referred to as "Cross Multiplication", was used to introduce the concept of Algebraic forms in Unit 2.

EXAMPLE:

$$\frac{5}{6} = \frac{15}{18} \quad T \quad F$$

Check by Cross Multiplication:
5 times 18 = 6 times 15 ? True

CONCEPT APPLICATIONS:

Set 1

True or False?

$\frac{1}{2} = \frac{7}{13}$ T F		$\frac{3}{4} = \frac{48}{64}$ T F	
$\frac{3}{4} = \frac{15}{20}$ T F		$\frac{5}{8} = \frac{20}{32}$ T F	
$\frac{7}{8} = \frac{60}{100}$ T F		$\frac{1}{4} = \frac{16}{32}$ T F	
$\frac{9}{16} = \frac{3}{8}$ T F		$\frac{3}{8} = \frac{12}{32}$ T F	
$\frac{7}{32} = \frac{15}{64}$ T F		$\frac{15}{16} = \frac{60}{64}$ T F	

Set 2

Solve the equations:

(1) $\dfrac{y}{2} = 2$; y = _____

(2) $\dfrac{r}{6} = 3$; r = _____

(3) $\dfrac{n}{10} = 5$; n = _____

(4) $\dfrac{s}{2} = 3$; s = _____

(5) $\dfrac{d}{5} = 6$; d = _____

(6) $\dfrac{n}{4} = 25$; n = _____

(7) $\dfrac{r}{8} = 32$; r = _____

(8) $\dfrac{x}{2} = 250$; x = _____

(9) $\dfrac{a}{3} = 75$; a = _____

(10) $\dfrac{d}{5} = 13$; d = _____

(11) $\dfrac{y}{16} = 4$; y = _____

(12) $\dfrac{a}{8} = \dfrac{1}{4}$; a = _____

(13) $\dfrac{d}{6} = \dfrac{1}{3}$; d = _____

(14) $\dfrac{m}{6} = \dfrac{1}{2}$; m = _____

(15) $\dfrac{r}{10} = \dfrac{1}{5}$; r = _____

(16) $\dfrac{-n}{12} = \dfrac{1}{3}$; n = _____

(17) $\dfrac{n}{12} = \dfrac{1}{4}$; n = _____

(18) $\dfrac{2n}{3} = 6$; n = _____

(19) $\dfrac{3m}{4} = 3$; m = _____

(20) $\dfrac{3x}{2} = 6$; x = _____

(21) $\dfrac{2d}{5} = 8$; d = _____

(22) $\dfrac{5r}{6} = 5$; r = _____

(23) $\dfrac{3y}{8} = 12$; y = _____

(24) $\dfrac{5c}{8} = 2\dfrac{1}{2}$; c = _____

Set 3

Solve the equations:

(1) $2a + 1 = 3$; $a =$ _____

(2) $4x + 5 = 9$; $x =$ _____

(3) $5x + 4 = 9$; $x =$ _____

(4) $3a - 8 = 13$; $a =$ _____

(5) $6d + 1 = 7$; $d =$ _____

(6) $\frac{3}{4}m = 6$; $m =$ _____

(7) $\frac{3}{7}y = 9$; $y =$ _____

(8) $\frac{3}{8}d = 12$; $d =$ _____

(9) $\frac{3}{2}x = 45$; $x =$ _____

(10) $\frac{x}{2} + 1 = 3$; $x =$ _____

(11) $\frac{a}{2} + 5 = 9$; $a =$ _____

(12) $\frac{x}{5} + 3 = 7$; $x =$ _____

(13) $\frac{x}{2} - 3 = 1$; $x =$ _____

(14) $\frac{x}{5} - 2 = 1$; $x =$ _____

(15) $5 = \frac{1}{5}x - 2$; $x =$ _____

(16) $0 = \frac{1}{4}x - 4$; $x =$ _____

(17) $6 = \frac{1}{3}x + 2$; $x =$ _____

(18) $6 = \frac{1}{2}x - 3$; $x =$ _____

Set 4

Solve the equations:

(1) $2x + 3 = 12 - x$; $x =$ _____

(2) $5x - 3 = 3x + 1$; $x =$ _____

(3) $4x - 5 = 4 + 3x$; $x =$ _____

(4) $3a + 6 = 14 - a$; $a =$ _____

(5) $8b + 7 = 3b + 12$; $b =$ _____

(6) $7x - 3 = 4x + 21$; $x =$ _____

(7) $8n - 7 = 5n + 14$; $n =$ _____

(8) $3x + 4 = x + 10$; $x =$ _____

(9) $9y - 4 = 4y + 6$; $y =$ _____

(10) $7b - 5 = 19 + b$; $b =$ _____

(11) $10m + 7 = 14 - 4m$; $m =$ _____

(12) $8x + 6 = 9 - 5x$; $x =$ _____

Set 5

Solve for the unknown:

(1) x + 9 = 17

(2) Y + 2 = 3.5

(3) M + 4.25 = 8.6

(4) 12 = 9 + X

(5) 7 = Y − 9

(6) $\frac{X}{5}$ = 10

(7) $\frac{X}{4}$ = 6

(8) $\frac{4}{X}$ = 6

(9) $\frac{Y}{2.5}$ = 4

(10) 4X = 24

(11) 36 = 6X

(12) Y − $\frac{1}{4}$ = 6$\frac{3}{4}$

(13) $\frac{1}{2}$ + X = 5.5

Set 6

Solve for:

(1) $5X = 26$ X _____

(2) $9 = \dfrac{72}{X}$ X _____

(3) $2\tfrac{1}{2} Y = 6\tfrac{1}{4}$ Y _____

(4) $5\tfrac{2}{3} + Y = 11\tfrac{1}{3}$ Y _____

(5) $8 = \dfrac{Y}{2\tfrac{1}{2}}$ Y _____

(6) $12 = Y - 3$ Y _____

(7) $\tfrac{2}{3} = X - 2$ X _____

(8) $Y + 6 = 16 - 9$ Y _____

(9) $25 + 5 = X - 3$ X _____

(10) $6\tfrac{1}{8} X = 7$ X _____

(11) $Y - 3 = 5 - Y$ Y _____

(12) $\dfrac{X}{3} = \dfrac{3}{X}$ X _____

UNIT 26

FORMULAS

INTRODUCTION

Formulas are equations having terms that express standard math rules. Solving formulas is done by rearranging the basic formula so that a single unknown quantity is isolated on one side of the equation and its equivalent terms are on the other side of the equal sign. The ability to apply the principles of equality to formulas enables the machinist to rearrange any formula so that he can easily solve for an unknown quantity.

OBJECTIVES:

After completing this unit the student will be able to:

- Solve an equation using the equality principle of powers or roots
- Solve a formula for any variable (letter) it contains

💡 EQUALITY PRINCIPLE OF POWERS

If both members (sides) of an equation are raised to the same power the sides will remain equal:

EXAMPLE 1:

$$\begin{aligned} \text{If} \quad 2 &= 2, \\ (2)^2 &= (2)^2 \quad \text{raise both sides to the second power} \\ 4 &= 4 \end{aligned}$$

EXAMPLE 2:

If	(x + 2)	=	3	then x = 1 is **true**
	(x + 2)²	=	3²	square both sides
	x + 4x + 4	=	9	evaluate: replace x with 1 and simplify
	(1) + 4(1) + 4	=	9	
	1 + 4 + 4	=	9	
	9	=	9	equation is still **true**

EXAMPLE 3:

If $\quad x = 7$

$\quad (x)^2 = (7)^2$

$\quad x^2 = 49$

💡 EQUALITY PRINCIPLE OF ROOTS

If the same root is taken of both sides of an equation the two sides will remain equal:

EXAMPLE 1:

If	9 · 16	=	144	is true, (take the square root of both sides)
then	√9 · 16	=	√144.	or
	√9 · √16		√144.	(simplify)
	3 · 4	=	12	or
	12	=	12	is still **true**

EXAMPLE 2:

If	(x + 2)²	=	5²	then
	√(x + 2)²	=	√5²	or
	x + 2	=	5	therefore
	x	=	3	now check by substituting x = 3 in the first equation.

EXAMPLE 3:

If	x²	=	100.
	√x²	=	√100.
then	x	=	10

167

SOLVING FORMULAS

The equality of the parts of a formula is not changed by using EQUALITY PRINCIPLES, however, the form will be changed. This rearrangement allows for the formula to be written in several equivalent forms which often make it easier to solve for a specific variable. Study the basic formulas that follow.

DISTANCE FORMULA ($d = rt$)

Distance Traveled = Rate of Speed x Time
(miles)   (mph)   (hours)

$$d = r \cdot t$$

What is the distance (d)?

$$d = 150 \text{ miles} = 15 \text{ mph for } 10 \text{ hrs, is true}$$

or

What is the rate of travel?

$$r = \frac{150 \text{ miles}}{10 \text{ hours}} = \frac{15 \text{ miles}}{1 \text{ hour}} = 15 \text{ mph}$$

or

How long does it take?

$$t = \frac{150 \text{ miles}}{15 \text{ mph}} = 10 \text{ hrs}$$

THEREFORE:

If $d = rt$ then we can find: $r = \frac{d}{t}$ and $t = \frac{d}{r}$

by applying the Division Principle as follows:

$$d = rt \qquad\qquad d = rt$$

$$\frac{d}{t} = \frac{rt}{t} \qquad\qquad \frac{d}{r} = \frac{rt}{r}$$

$$\frac{d}{t} = r \qquad\qquad \frac{d}{r} = t$$

AREA OF RECTANGLE FORMULA ($A = \ell w$)

If $A = \ell w$ or $A = \ell w$

$\dfrac{A}{\ell} = \dfrac{\ell w}{\ell}$ $\dfrac{A}{w} = \dfrac{\ell w}{w}$

$\dfrac{A}{\ell} = w$ $\dfrac{A}{w} = \ell$

EQUIVALENT FORMS

Let $10'' = \ell$ $A = 70$ sq. in. $7'' = w$

$A = \ell w$ or $\ell = \dfrac{A}{w}$ or $w = \dfrac{A}{\ell}$

$70 = 10 \cdot 7$ $10 = \dfrac{70}{7}$ $7 = \dfrac{70}{10}$

$70'' = 70''$ $10'' = 10''$ $7'' = 7''$

AREA OF A CIRCLE FORMULA ($A = \pi r^2$).

This formula can be rearranged to an equivalent form such that the radius (r) can be determined.

$$A = \pi r^2$$
$$\dfrac{A}{\pi} = \dfrac{\pi r^2}{\pi}$$
$$\dfrac{A}{\pi} = r^2$$
$$\sqrt{\dfrac{A}{\pi}} = \sqrt{r^2}$$
$$\sqrt{\dfrac{A}{\pi}} = r$$

PROCEDURE FOR SOLVING FORMULAS:

Any formula can be solved by applying the proper PRINCIPLE OF EQUALITY for addition, subtraction, multiplication, division, powers and roots. The rules for ORDER OF OPERATION must also be carefully observed. The following table summarizes all the skills you have previously learned that are required for solving formulas:

THE GOLDEN RULE OF EQUATIONS:

Any operation (such as: $+$, $-$, $\times$, $\div$, $(\ )^2$ or $\sqrt{\ }$) may be performed on one side of an equation using any number if you perform the same operation on the other side with the same number.

ORDER OF OPERATIONS LAW:

Step 1 Remove symbols of inclusion starting with the innermost and complete the operations on terms they contain.

Step 2 Simplify indicated powers and roots where possible.

Step 3 Perform the indicated multiplication and division order from left to right.

Step 4 Perform the addition and subtraction in order from left to right.

CONCEPT APPLICATIONS:

Set 1   Solve for:

(1) $A = LW$ $L = \underline{\quad}$ $W = \underline{\quad}$

(2) $P = 2(L + W)$ $W = \underline{\quad}$ $L = \underline{\quad}$

(3) $A = \dfrac{BH}{2}$ $B = \underline{\quad}$ $H = \underline{\quad}$

(4) $S.F.M = \dfrac{(D)(R.P.M.)}{4}$ $D = \underline{\quad}$ $R.P.M. \underline{\quad}$

      Note: S.F.M. is a single term meaning surface feet per minute

(5) $T.P.F. = \dfrac{LD - SD}{L} \times 12$ $L = \underline{\quad}$ $SD = \underline{\quad}$ $LD = \underline{\quad}$

      Note: T.P.F. is a term meaning taper per foot

(6) $A = \dfrac{1}{D.P.}$ $D.P. = \underline{\quad}$

(7) $P.D. = \dfrac{N}{D.P.}$ $N = \underline{\quad}$ $D.P. \underline{\quad}$

(8) $V = \dfrac{LWH}{3}$ $L = \underline{\quad}$ $W = \underline{\quad}$ $H = \underline{\quad}$

(9) $\dfrac{2}{3} = X - 2$ $X = \underline{\quad}$

(10) $Y + 16 = 16 - 9$ $Y = \underline{\quad}$

(11) $C = D$ $D = \underline{\quad}$

(12) $\dfrac{D}{L} = \dfrac{T}{12}$ T = _____ L = _____

(13) $D = \dfrac{N + 2}{P}$ N = _____ P = _____

Set 2

(1) P = 0.2w + 0.3 w = _____

(2) Sin = $\dfrac{\text{Opp.}}{\text{Hyp.}}$ Opp.= _____ Hyp.= _____

(3) Cos = $\dfrac{\text{Adj.}}{\text{Hyp.}}$ Adj.= _____ Hyp.= _____

(4) Tan = $\dfrac{\text{Opp.}}{\text{Adj.}}$ Opp.= _____ Adj.= _____

(5) Csc = $\dfrac{\text{Hyp.}}{\text{Opp.}}$ Hyp.= _____ Opp.= _____

(6) Sec = $\dfrac{\text{Hyp.}}{\text{Adj.}}$ Hyp.= _____ Adj.= _____

(7) Cot = $\dfrac{\text{Adj.}}{\text{Opp.}}$ Adj.= _____ Opp.= _____

Set 3

(1) $A = \dfrac{(a + b)h}{2}$ a = ___ b = ___ h = ___

(2) $A = \pi r^2$ r = _____

(3) $F = \dfrac{1}{2} D (1 + \tan B)$ tan B = _____

(4) $A = \pi^2 \dfrac{\theta}{360}$ $\theta =$ _____

(5) $P = \dfrac{L}{N}$ $L =$ _____ $N =$ _____

(6) $T = \dfrac{P}{2} \cos C$ $P =$ _____ $\cos C$ _____

(7) $D = Pd + 2a$ $Pd =$ _____

(8) $Pd = \dfrac{N}{P \cos C}$ $P =$ _____ $N =$ _____ $\cos C$ _____

(9) $0 = \dfrac{T \times L}{2}$ $T =$ _____ $L =$ _____

(10) $\dfrac{L}{\ell} = \dfrac{d}{T}$ $d =$ _____ $T =$ _____

(11) $\% = \dfrac{(D - d)\,100}{L}$ $D =$ ____ $d =$ ____ $L =$ ____

(12) $c = \dfrac{P}{z(\sin B)}$ $\sin B =$ _____

(13) $u = V\left(\dfrac{C - W}{C}\right)$ $W =$ _____ $V =$ _____

Set 4

(1) $F = \frac{9}{5}C + 32$ $C = \underline{\hspace{1cm}}$

(2) $A = P + .06\,Pt$ $t = \underline{\hspace{1cm}}$ $P = \underline{\hspace{1cm}}$

(3) $F = \frac{w}{g}a$ $a = \underline{\hspace{1cm}}$ $g = \underline{\hspace{1cm}}$

(4) $V = 1090 + 1.14(t - 32)$ $t = \underline{\hspace{1cm}}$

(5) $1 = \frac{\ell}{mx}$ $x = \underline{\hspace{1cm}}$ $\ell = \underline{\hspace{1cm}}$

(6) $T = \frac{s}{2} + 3$ $s = \underline{\hspace{1cm}}$

Set 5

(1) If 15 men can do a piece of work in 28 days, in how many days can 12 men do the same piece of work?

(2) If a bar of steel 6 ft. long weighs 380 lbs., how much would the same bar weigh if it were 5 ft. long?

(3) If 8 men can machine 336 castings in one day, how many castings would 11 men machine in one day?

(4) If an airplane flies 1,950 miles in 8 hours, how far would it fly in 10 hours?

(5) If a driving pulley 22 inches in diameter runs at a speed of 175 R.P.M., what will be the speed of a 14 inch diameter driven pulley?

GEOMETRY

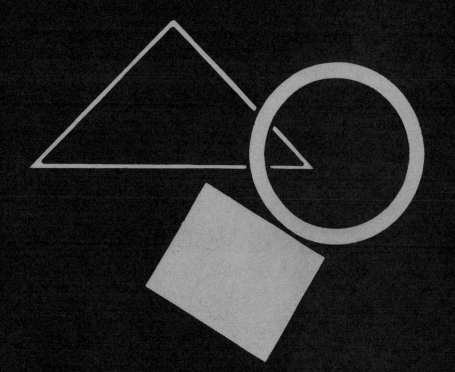

GEOMETRY SYMBOLS

ANGLE	∠		THEREFORE	∴
PARALLEL	∥		ARC	⌒
PERPENDICULAR	⊥		EQUAL	=
TRIANGLE	△		PI	π
RIGHT TRIANGLE	◣		CIRCLE	⊙
RIGHT ANGLE	∟		RADIUS	R, r
			CIRCUMFERENCE	C
DEGREE	°		LINE	←•——•→
MINUTE	′		LINE SEGMENT	A———B
SECOND	″		INTERSECTION	∩

UNIT 27

AXIOMS AND PROPOSITIONS

INTRODUCTION

Geometry is the study of the properties, measurements and relationships of points, lines, angles, surfaces and solids. The process for solving geometry problems applies accepted facts about points, lines and planes as building blocks to establish additional facts and reach logical conclusions. Certain axioms and propositions that have been proven or accepted are the facts to start with. Geometry is applied to machine shop problems to make the calculations from engineering drawings that precede the making of machined parts. The processes used in solving the problems in this text are the same processes needed to plan and layout machine parts.

OBJECTIVES:

After completing this unit the student will be able to:

- Define axiom and proposition
- Define the basic terms of geometry
- Illustrate geometric axioms with drawings

DEFINITIONS:

AXIOM

A rule which is accepted as true without further proof. (Also called **postulates**)

PROPOSITION

A rule which has been proven true. (Also called **theorems**)

✋ GEOMETRY

The study of points, lines, and planes in space. POINTS, LINES and PLANES have no definitions, but are used in a general way to help define other terms. They are frequently related in a special way which has many shop implications for problem solving. Understanding these relations and their applications depends on our accepting given definitions, axioms and propositions as TRUE.

BASIC TERMS: GENERAL USE AND CHARACTERISTICS

💡 POINT

A **point** is named with a capital letter (Point A). A **point** is used to identify a specific position or location on a line, in a plane or in space.

EXAMPLE 1:   End **points** of a segment: (A and B)

EXAMPLE 2:   Intersection **point(s)** of lines: (A)

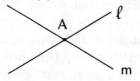

EXAMPLE 3:   Center **point** (origin) of a circle: (O)

EXAMPLE 4:   Every segment has exactly 1 **midpoint**.
A midpoint always divides the segment into two equal parts.

```
        10
A •————•————• B      AM  =  MB
        M              5  =   5
                 and AB  =  5 + 5  =  10
```

💡 LINE

A **line** contains a set of continuous points which extend infinitely in opposite directions.

A **line** means a straight line unless otherwise designated.

A **line** is named by one small letter (line ℓ) or with two capital letters which represent any two points on the line (Line AB). (The length of line segment from point A to point B is written AB.

EXAMPLE 1:   Perpendicular lines: (⊥) or (⊥)

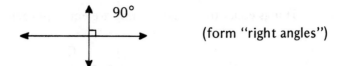

(form "right angles")

EXAMPLE 2:   Parallel lines: (||)

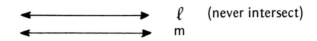

(never intersect)

EXAMPLE 3:   Intersecting lines:

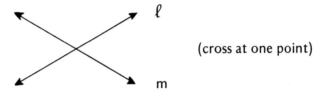

(cross at one point)

PLANE

A **plane** is a continuous flat surface extending infinitely.

A **plane** is named with a capital letter.

EXAMPLE 1:   Parallel planes do not intersect, such as a floor and ceiling.

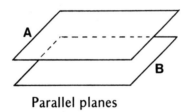

Parallel planes

EXAMPLE 2:   Perpendicular planes intersect in a straight line (ℓ).

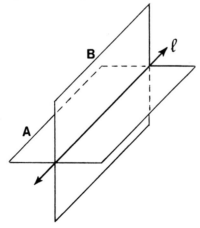

Perpendicular planes

179

BASIC AXIOMS:

💡 1. Things equal to the same thing are equal to each other.

If A •———• B (5″) = C •———• D (5″) and

If X •———• Y (5″) = C •———• D (5″) then AB = XY

💡 2. Any number or quantity may be substituted for its equal.

If a = (X + Y) and if (X + Y) = 10
then a = 10

💡 3. If equals are added to equals, the results are equals.

If A •——5″——• B and C •——2″——• D   If   a = b and
                                                  c = d
    P •——5″——• Q and R •——2″——• S   Then a + c = b + d

then AB + CD = PQ + RS since
      5 + 2 = 5 + 2

💡 4. If equals are subtracted from equals, the results are equals.

If line segments AB and CD each equal 10 inches, and you remove 2 inches from each one, the remaining segments have equal lengths of 8 inches each.

or   A •——10″——/2″/• B   C •——10″——/2″/• D

💡 5. If equals are multiplied by equals, the results are equal.

$$\begin{array}{rcl} 20 & = & 20 \\ 20 \cdot 5 & = & 20 \cdot 5 \\ 100 & = & 100 \end{array} \qquad \begin{array}{rcl} a & = & b \\ a \cdot c & = & b \cdot c \\ ac & = & bc \end{array}$$

6. If equals are divided by equals, the results are equals.

$$60 = 60$$
$$60 \div 5 = 60 \div 5$$
$$12 = 12$$

$$a = b$$
$$\frac{a}{c} = \frac{b}{c}$$

7. A whole unit is greater than any of its parts and is equal to the sum of its parts.

$$12'' = 6'' + 4'' + 2''.$$

12 is greater than 6, 4 or 2, but is equal to their sum.

8. Two points fix the position of a line. Only one line can be drawn through 2 given points.

Fix points A and B. Only one line can pass through or contain these points.

9. Two straight lines in the same plane can only intersect in 1 point: (at most)

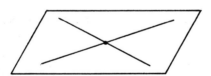

10. Given a line, ℓ and a point, P, which is not on ℓ, there exists exactly one line m, passing through P which is also parallel to the given line ℓ.

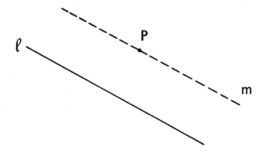

181

UNIT 28

ANGLES AND LINES

INTRODUCTION

The first thing to learn about geometry is the various ways in which lines can be drawn, angles formed, and the interrelationship that results. The angle is formed when two lines, called legs, intersect. We recognize an angle as being measured in degrees, minutes and seconds.

OBJECTIVES:

After completing this unit the student will be able to:

- Define parts (subsets) of a line
- Identify perpendicular (⊥) lines and segments
- Identify parallel (||) lines and segments
- Distinguish between types of angles (using measures)
- Determine measures for complementary and supplementary angles
- Determine angle measure expressed in degrees, minutes and seconds

LINE VS SEGMENT:

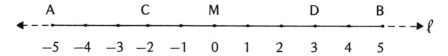

182

LINE

The continuous set of points extending (without end) in opposite directions is called a line. It can be named using any 2 points, such as:

Line AB ($\overleftrightarrow{AB}$) = Line CM ($\overleftrightarrow{CM}$) = Line MD ($\overleftrightarrow{MD}$)

or with the single letter, ℓ, or as line ℓ.

NOTE:

The line cannot be measured since the points (or length) never stop.

SEGMENT

The set of any two points (end points) on a line and all points between them:

The segment (or line segment) is symbolized as ($\overline{AB}$). The measure of a segment is given with a number to represent its length. Segment AB is given a measure of "7", since the end points are identified as "3" and "10". (The length of a segment is equal to the difference in the numbers representing the end points.) Segment AB is also equal to segment BA, or ($\overline{AB}$) = ($\overline{BA}$). The measure of segment AB is symbolized as (AB).

COMPARE:

($\overleftrightarrow{AB}$) means **line** AB.

($\overline{AB}$) means **segment** AB.

(AB) means the **measure** (number) of AB.

NOTE:

In the study of geometry, a line (ℓ or $\overleftrightarrow{AB}$) and a line segment ($\overline{AB}$ or $\overline{BA}$) have separate and distinct meanings. However, in the shop both are commonly called "a line". We usually hear, "The measure of the **line** is 7″ when we really mean "The measure of that **line segment** is 7″. The distinction, or lack of distinction, will not cause you any confusion as you progress from the study of math theory and begin to use practical math in machine shop applications.

✋ RAY

A set of points on a line extending in only one direction. (also called a half line):

The ray AB is symbolized as, ($\overrightarrow{AB}$)

NOTE:

$\overrightarrow{AB} \neq \overrightarrow{BA}$ since the first letter is used to represent the starting point and the second letter gives the direction.

✋ PERPENDICULAR

Two lines that meet to form right angles, (rt. ∠ s) are said to be perpendicular to each other. Line segments and rays contained in perpendicular lines are also perpendicular:

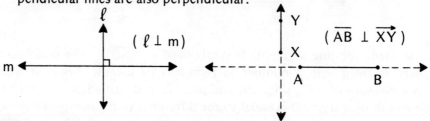

✋ PARALLEL

Two or more lines in the same plane that never meet (or intersect), are said to be parallel. Line segments and rays contained in parallel lines are also parallel:

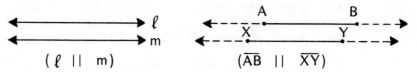

✋ ANGLE

An angle (∠) is the figure formed by joining two rays at the same point (origin or vertex): The rays are called sides.

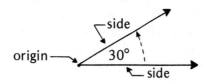

The angle may be named in several ways:

(1)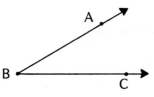

∠ ABC   or   ∠ CBA
with middle letter as
as the vertex pt.

(2)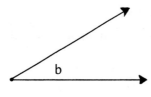

∠ b   a small letter
(inside the angle)

(3)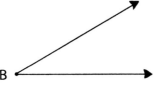

∠ B   a capital letter
(outside the vertex);
used only for 1 ∠

(4)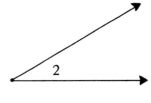

∠ 2   a number

NOTE:

Numbering ∠s is very useful in drawings where several ∠s must be identified:

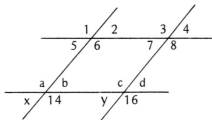

Angles are measured in degrees:

    1 degree = $\frac{1}{360}$ of a circle   (°)   Just as hours

    1 minute = $\frac{1}{60}$ of 1 degree   (′)   minutes

    1 second = $\frac{1}{60}$ of 1 minute   (″)   seconds

                                                                 relate to the measure of time.

OBSERVE:

Twenty-five degrees, fourteen minutes, and thirty-seven seconds is written: 25° 14′ 37″

TYPES OF ANGLES (by measure)

✋ ACUTE ANGLE
An ∠ with measure greater than zero° and less than 90°:

(0° < m < 90°) means the measure **between** 0° and 90°.

✋ RIGHT ANGLE
An ∠ with measure of exactly 90 degrees.

(rt. ∠ = 90°)   rt. ∠ symbol is →

✋ OBTUSE ANGLE
An ∠ with measure greater than 90° and less than 180°:

(90° < m < 180°)

✋ STRAIGHT ANGLE
An ∠ with measure of exactly 180 degrees.

(st. ∠ = 180°)

✋ REFLEX ANGLE
An ∠ with measure greater than 180° and less than 360°:

(180° < m < 360°)

✋ REVOLUTION
One revolution = 360°.

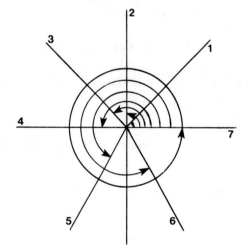

Identify each type of ∠ below:

1. _____
2. _____
3. _____
4. _____
5. _____
6. _____
7. _____

186

TYPES OF ANGLES   (by relative position)

✋ ADJACENT

Two ∠s that share one of their sides, but no interior points are said to be adjacent.

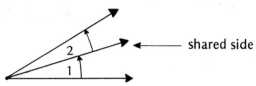
— shared side

✋ COMPLEMENTARY

Two ∠s whose sum = 90°. (or make a rt. ∠.)

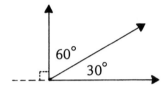   30° + 60° = 90° = rt.∠ .

✋ SUPPLEMENTARY

Two ∠s whose sum = 180°. (or make a st. ∠.)

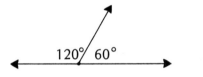   60° + 120° = 180° = st. ∠ .

CONCEPT APPLICATIONS

(1) Determine Angle A

Angle A = _____

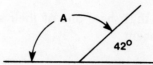

(2) Determine Angle B

Angle B = _____

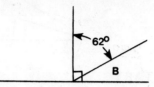

(3) Determine Angle C

Angle C = _____

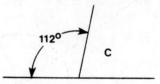

(4) List the supplementary angles for the following:

28 degrees _____   108 degree _____

106 degrees _____   12 degrees _____

120 degrees _____   176 degrees _____

92 degrees _____   7 degrees _____

46 degrees _____   135 degrees _____

45 degrees _____   27 degrees _____

30 degrees _____

(5) List the complementary angles for the following:

30 degrees _____   78 degrees _____

20 degrees _____   36 degrees _____

80 degrees _____   45 degrees _____

60 degrees _____   2 degrees _____

26 degrees _____   32 degrees _____

14 degrees _____   57 degrees _____

(6)

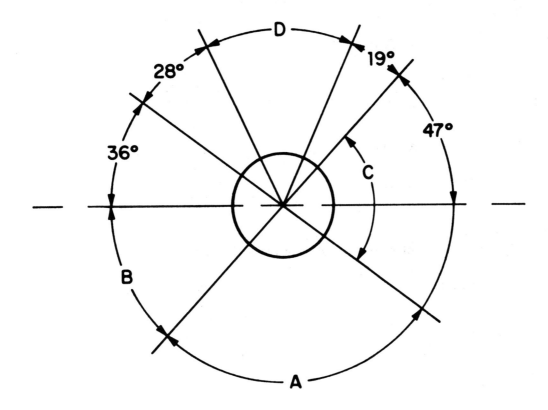

1. Determine ∠D = _____

2. Determine ∠A = _____

3. Determine ∠C = _____

4. Determine ∠B = _____

5. What is the complement of ∠D? _____

6. What is the supplement of ∠A? _____

7. What is the complement of ∠C? _____

8. What is the complement of ∠B? _____

UNIT 29

INTERSECTING LINES AND ANGLES

INTRODUCTION

The machinist often must calculate the relationship between two or more points on a workpiece. This is done by drawing lines between them and forming angles. Then the machinist determines which of the relationships will help measure the distances and the angle relationships between these points. This process may be repeated a number of times on a single blueprint to determine how to do all of the machining required to complete the part. This information may be fed into a computer for use on a computer numerical control (CNC) machine tool or it may be used in machine setup for manual machining operations on a manually operated machine.

OBJECTIVES:

After completing this unit the student will be able to:

- Make observations about given information and state conclusions
- Determine angle measures using
    a. Vertical angles
    b. Parallel Lines and Transversals
    c. Exterior Angles

ANGLE MEASURES

Are numbers and can be added together:

EXAMPLES:

$$\angle 1 + \angle 2 = 180°$$
the measure is: $40° + 140° = 180°$

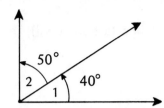

$\angle 1 + \angle 2 = 90°$
the measure is: $40° + 50° = 90°$

💡 PROPOSITION I:

If two straight lines intersect, the opposite or vertical angles are equal.

Angles formed by two intersecting lines are called **vertical** angles or **opposite** angles. $\angle 1$ and $\angle 2$ are vertical angles.

OBSERVATIONS:

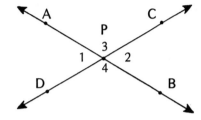

$\overleftrightarrow{AB} \cap \overleftrightarrow{CD}$ at point P

("$\cap$" means — intersection)

$\angle 1 = \angle 2$ since vertical $\angle$s are = (Prop. I)
$\angle 3 = \angle 4$ again vertical $\angle$s are = (Prop. I)

$\angle 1 + \angle 3 = 180°$ definition of supplementary $\angle$s
$\therefore \angle 3 + \angle 2 = 180°$
and $\angle 2 + \angle 4 = 180°$
$\angle 4 + \angle 1 = 180°$

NOTE:

"$\therefore$" means "then" or "therefore" and is frequently used.

💡 PROPOSITION II:

If two or more lines in the same plane are perpendicular to the same line, they are parallel to each other.

Line ℓ and line m are perpendicular to line n.

OBSERVATIONS:

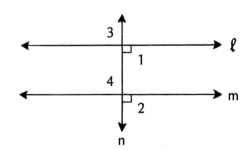

$\ell \perp n$
$m \perp n$
$\therefore \ell \parallel m$

$\therefore \angle 1 = \angle 2 = \angle 3 = \angle 4 = $ rt. $\angle$s $= 90°$

191

PROPOSITION III:

If two parallel lines are cut by a third line (a transversal), then the pairs of alternate interior angles are equal.

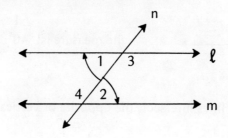

∠1 = ∠2   since alternate interior ∠s = (Prop. III)

∴ ∠3 = ∠4   again alternate interior ∠s = (Prop. III)

NOTE:

If we combine the Proposition I that vertical ∠s are equal and Proposition III that alternate interior ∠s are equal, a new series of observations can be seen:

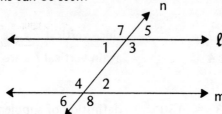

PROPOSITION IV:

If two parallel lines are cut by a third line (a transversal), then the pairs of alternate exterior angles are equal.

∠5 and ∠6 are alternate exterior angles

OBSERVATIONS:

∠5 = ∠6   combine alternate interior angles (Prop. III)
∠7 = ∠8   and vertical angles (Prop. I)

PROPOSITION V:

If two parallel lines are cut by a third line (a transversal), then the pairs of corresponding angles are equal.

∠2 and ∠5 are corresponding angles

OBSERVATIONS:

∠1 = ∠6   ∠2 = ∠5   ∠3 = ∠8   ∠4 = ∠7

PROPOSITION VI: The sum of the interior angles of a triangle is 180°.

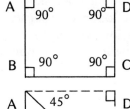

A square has 4 rt. ∠s (= 90°)
The sum of 4 rt ∠s is 4 · 90° = 360°.

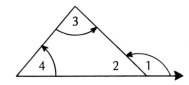

However, if the square is divided into two triangles, then △ ABC has angles of: 45° + 45° + 90° = 180°.
and ∠ABC + ∠BAC + ∠BCA = 180°

PROPOSITION VII:
An exterior angle of a triangle is equal to the sum of the two interior angles at the other vertices.

∠1 is an Exterior Angle. How does ∠1 relate to ∠3 and ∠4?

OBSERVATIONS:

∠2 + ∠3 + ∠4 = 180°   Interior ∠s of △ = 180° (Prop.VI)
∠1 + ∠2 = 180°   Supplementary ∠s = 180°
∴ ∠1 + ∠2 = ∠2 + ∠3 + ∠4   by substitution (Axiom 1)
and ∠1 = ∠3 + ∠4   by subtraction (Axiom 4)

PROBLEM SOLVING

Solving problems in geometry requires the use of algebraic skills in applying the geometric concepts you learned as basic axioms, propositions and definitions. Since a number of steps are often involved in finding a final solution, each step should be supported by a **basic axiom, proposition or definition.**

EXAMPLE:

Given: △ ABC is a rt. ∠ ,
$\angle 3 = 47°$

Find: $\angle 1 =$?

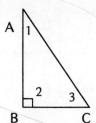

Steps	Supporting reason:
1. $\angle 1 + \angle 2 + \angle 3 = 180°$	Why? (Prop. VI)
2. $\angle 2 = 90°$	Why? (Given)
3. $\angle 1 + \angle 3 = 90°$	Why? (Axiom 4 step 1 - step 2)
4. $\angle 1 + 47° = 90°$	Why? (Axiom 2 applied to step 3)
5. $\angle 1 = 43°$	Why? (Axiom 4 applied to step 4)
or (simplified)	
1. $\angle 1 + 90° + 47° = 180°$	Why?
2. $\angle 1 + 137° = 180°$	Why?
3. $\angle 1 = 43°$	Why?

CONCEPT APPLICATIONS

Problem 1

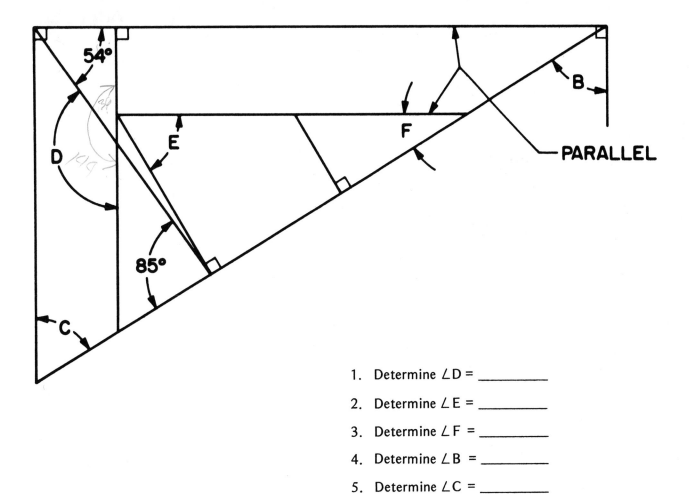

1. Determine ∠D = _____
2. Determine ∠E = _____
3. Determine ∠F = _____
4. Determine ∠B = _____
5. Determine ∠C = _____

Problem 2

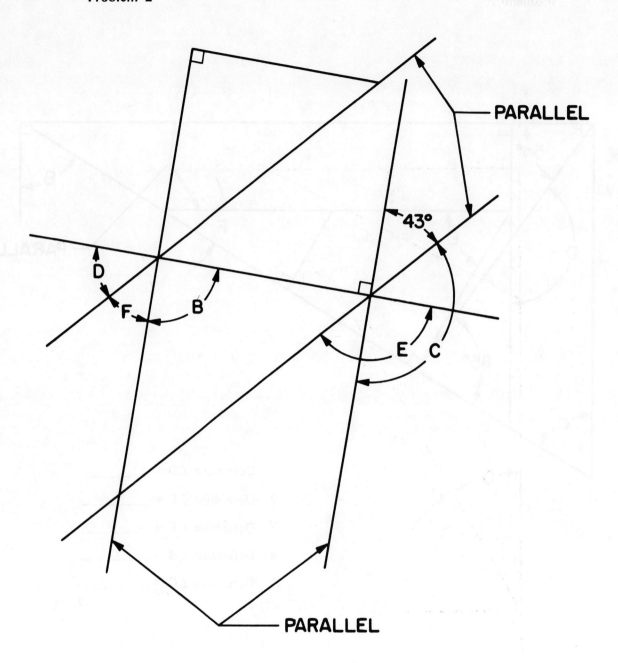

1. Determine ∠D = _____
2. Determine ∠E = _____
3. Determine ∠F = _____
4. Determine ∠B = _____
5. Determine ∠C = _____

Problem 3

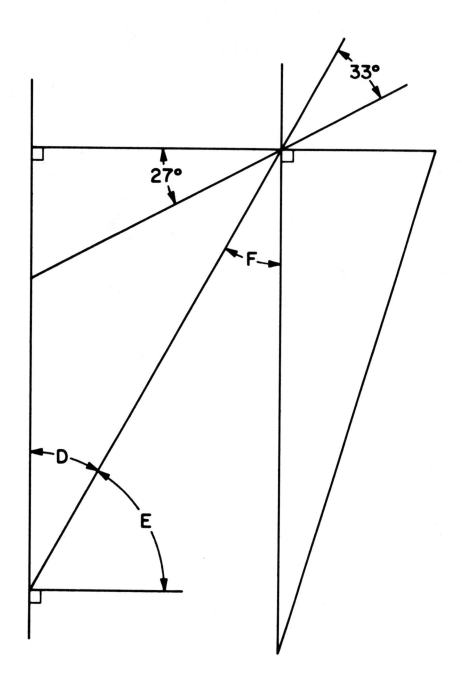

1. Determine ∠D = _____
2. Determine ∠E = _____
3. Determine ∠F = _____

Problem 4

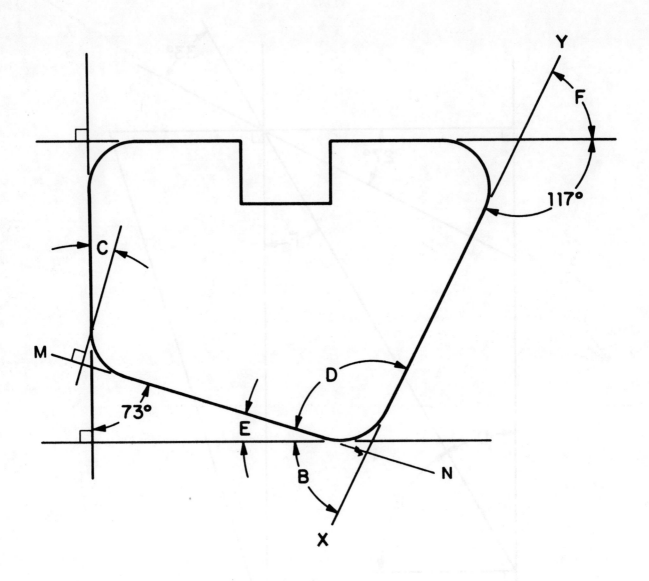

It is required to mill surface XY and NM. To position the block properly it is necessary to determine angle F and E.

Determine ∠E = _____

Determine ∠F = _____

Determine ∠C = _____

Problem 5

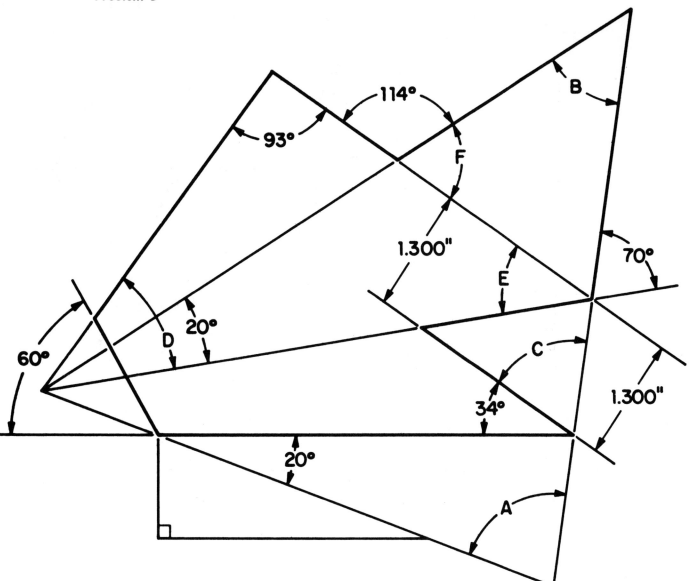

1. Determine ∠D = _____
2. Determine ∠E = _____
3. Determine ∠F = _____
4. Determine ∠B = _____
5. Determine ∠C = _____
6. Determine ∠A = _____

199

Problem 6

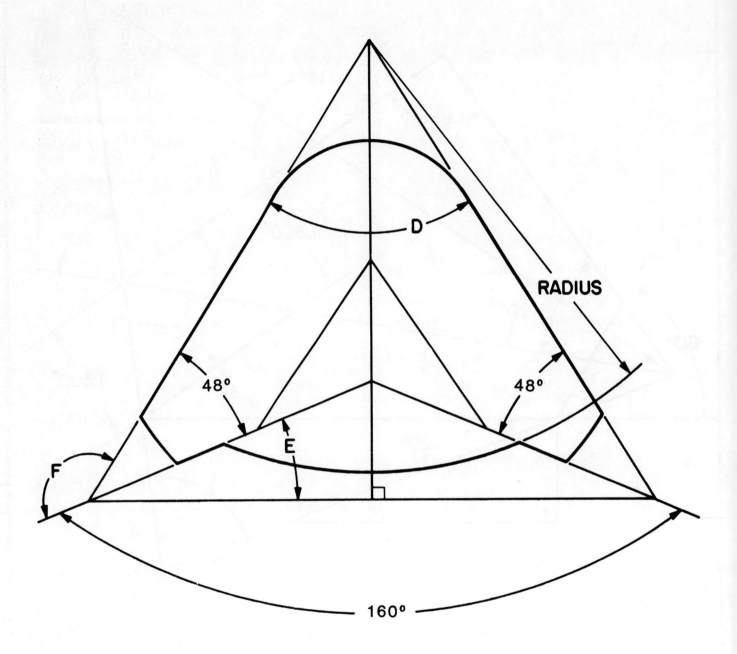

1. Determine ∠D = _____
2. Determine ∠E = _____
3. Determine ∠F = _____

UNIT 30

RELATING LINES AND ANGLES TO POLYGONS AND CIRCLES

INTRODUCTION

Most machining is done in straight lines. Therefore, it is important to position the workpiece on the table so that the machining can be done with a minimum number of setup changes. Also, the relation between lines and circles and their intersection is basic to measuring distances between holes, the distance across the dovetail slide or measurement of thread diameters.

OBJECTIVES:

After completing this unit the student will be able to:

- Identify properties of basic polygons
- Determine the sum of the interior angles of any polygon
- Determine equal angles using parallel and perpendicular sides of polygons.
- Use tangent lines to solve problems

TRIANGLES:

Triangles are 3 sided polygons.

BASIC TYPES:

RIGHT TRIANGLE

Any triangle containing a right (90°) angle.

201

✋ ISOSCELES TRIANGLE

A triangle having a pair of equal sides.

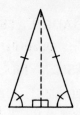

OTHER PROPERTIES:

1. ∠s opposite the equal sides are equal. (base ∠s)
2. The altitude (⊥) divides the △ into 2 congruent (≅) △s having their corresponding parts (sides and ∠s equal).

✋ EQUILATERAL TRIANGLE

A triangle having three equal sides.

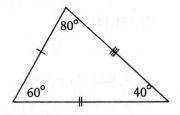

OTHER PROPERTIES:

1. 3 equal ∠s of 60 each (180° ÷ 3)
2. Altitude divides the △ into 2 congruent triangles, having all corresponding parts equal.

✋ SCALENE TRIANGLE

A triangle having three unequal sides.

OTHER PROPERTIES:

1. 3 unequal (≠) ∠s

✋ QUADRILATERALS:

Quadrilaterals are four sided polygons.

BASIC TYPES:

✋ SQUARE

A quadrilateral with:

1. 4 = sides, and
2. 4 = ∠s of 90°.

OTHER PROPERTIES:

1. Diagonals bisect the ∠s.
2. Diagonals bisect each other.
3. Diagonals form ≅ .
4. Opposite sides are ||.
5. Diagonals are ⊥ to each other.
6. Diagonals are =.

✋ RECTANGLE

A quadrilateral with:

1. Opposite sides equal, and
2. 4 = ∠s of 90°.

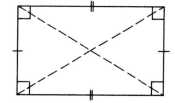

OTHER PROPERTIES:

1. Opposite sides ||.
2. Diagonals are =.
3. Diagonals bisect each other. (divide into equal parts)
4. Diagonals form pairs of ≅ △.

✋ PARALLELOGRAM

A quadrilateral with:

1. Opposite sides ||.

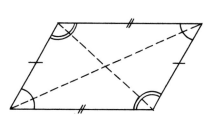

OTHER PROPERTIES:

1. Opposite sides =.
2. Opposite ∠s =.
3. Diagonals are ≠.
4. Diagonals bisect each other.
5. Diagonals form pairs of ≅ △.

203

PROPOSITION VIII:
If two angles have their corresponding sides parallel, then the angles are equal.

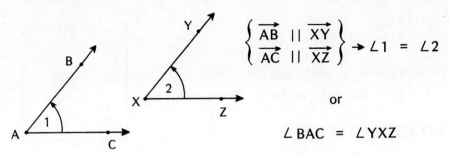

$$\left\{\begin{array}{c}\vec{AB} \parallel \vec{XY} \\ \vec{AC} \parallel \vec{XZ}\end{array}\right\} \rightarrow \angle 1 = \angle 2$$

or

$$\angle BAC = \angle YXZ$$

PROPOSITION IX:
If two angles have their corresponding sides perpendicular, then the angles are equal.

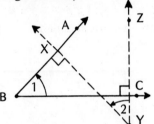

$$\left\{\begin{array}{c}\vec{BA} \perp \vec{YX} \\ \vec{BC} \perp \vec{YZ}\end{array}\right\} \rightarrow \angle 1 = \angle 2$$

PROPOSITION X:
The sum of the interior angles of a polygon is equal to 180° times the number of sides less 2 [or $S = 180° (n - 2)$].

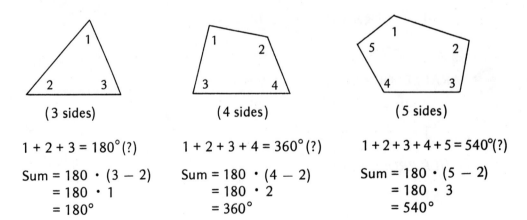

(3 sides)   (4 sides)   (5 sides)

1 + 2 + 3 = 180°(?)   1 + 2 + 3 + 4 = 360°(?)   1 + 2 + 3 + 4 + 5 = 540°(?)

Sum = 180 · (3 − 2)   Sum = 180 · (4 − 2)   Sum = 180 · (5 − 2)
      = 180 · 1   = 180 · 2   = 180 · 3
      = 180°   = 360°   = 540°

PROPOSITION XI

A line segment drawn from a point of tangency to the center of a circle is perpendicular to the tangent.

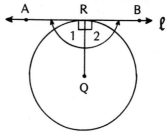

1. Line $\ell \cap \odot = R$
   (The intersection of line ℓ with circle Q is point R)

2. Point R is called the tangent point or point of tangency.

3. Line AB = line ℓ

4. Line $\ell \perp \overline{QR}$

5. $\angle 1 = \angle 2 = 90°$

PROPOSITION XII:

If two tangents are drawn to a circle from the same exterior point, then the corresponding segments are equal and a line from this external point to the center of the circle forms two equal angles.

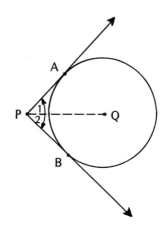

OBSERVATIONS:

1. $\overline{AP} = \overline{BP}$
2. $\overline{PQ}$ bisects $\angle APB$
3. $\angle 1 = \angle 2$ or

   $\therefore \angle APQ = \angle BPQ$

CONCEPT APPLICATIONS:

Problem 1

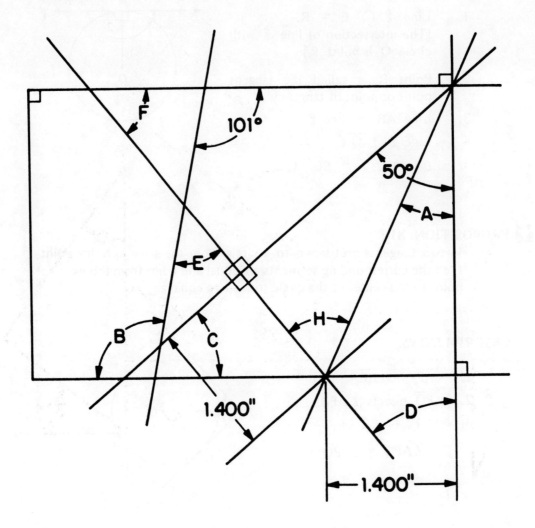

It is required of the machinist to see that the 1.400" dimensions given are the radii of an imaginary circle. Once that circle is constructed, the propositions of tangent lines apply.

1. Determine ∠D = _____
2. Determine ∠E = _____
3. Determine ∠F = _____
4. Determine ∠B = _____
5. Determine ∠C = _____
6. Determine ∠A = _____
7. Determine ∠H = _____

206

Problem 2

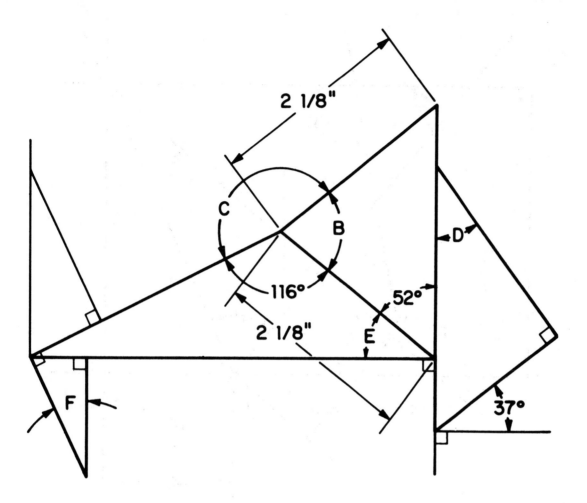

The $2\frac{1}{8}$ dimension is the key!

1. Determine ∠D = _____
2. Determine ∠E = _____
3. Determine ∠F = _____
4. Determine ∠B = _____
5. Determine ∠C = _____

Problem 3

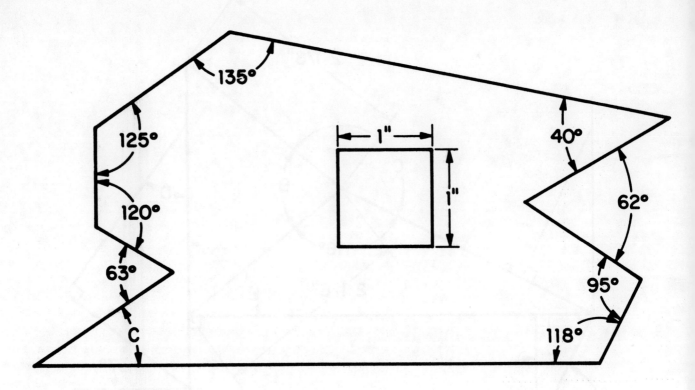

1. Determine ∠C = _____

Problem 4

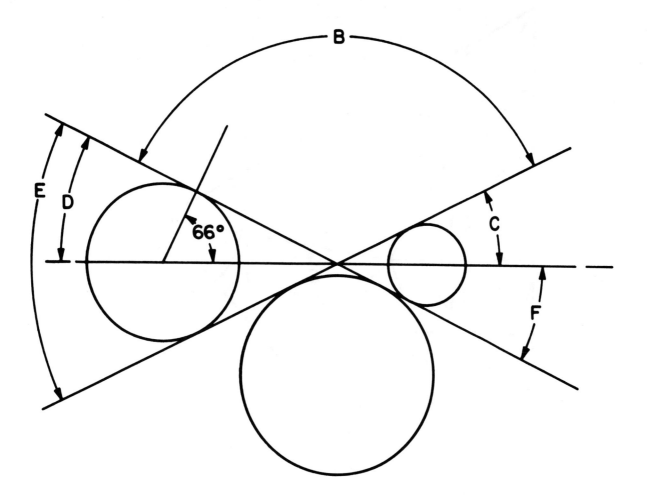

1. Determine ∠D = _____
2. Determine ∠E = _____
3. Determine ∠F = _____
4. Determine ∠B = _____
5. Determine ∠C = _____

Problem 5

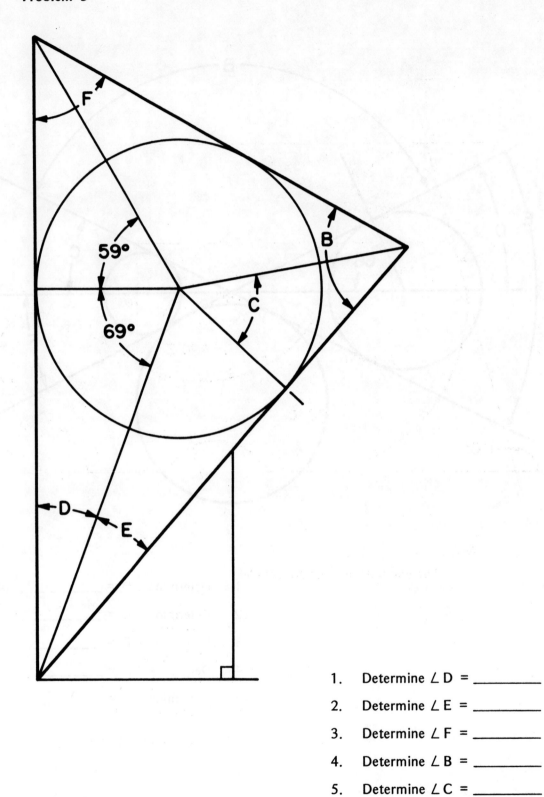

1. Determine ∠D = _____
2. Determine ∠E = _____
3. Determine ∠F = _____
4. Determine ∠B = _____
5. Determine ∠C = _____

Problem 6

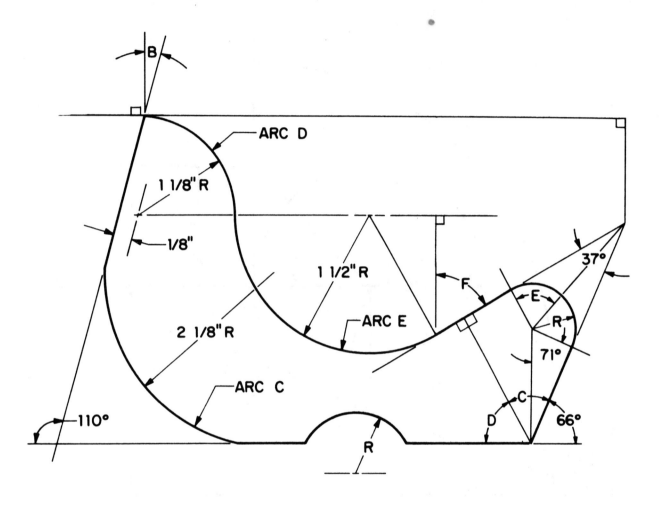

NOTE:
Top and bottom lines are parallel.

1. Determine ∠D = _____
2. Determine ∠E = _____
3. Determine ∠F = _____
4. Determine ∠B = _____
5. Determine ∠C = _____

211

Problem 7

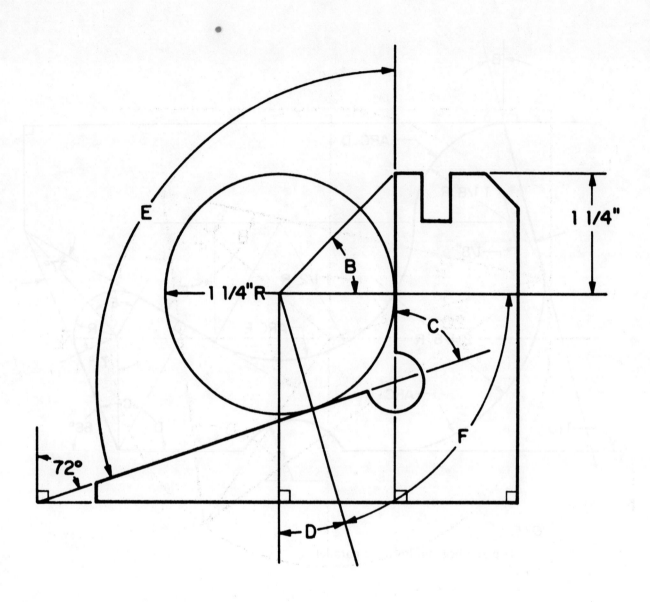

1. Determine ∠D = _____
2. Determine ∠E = _____
3. Determine ∠F = _____
4. Determine ∠B = _____
5. Determine ∠C = _____

Problem 8

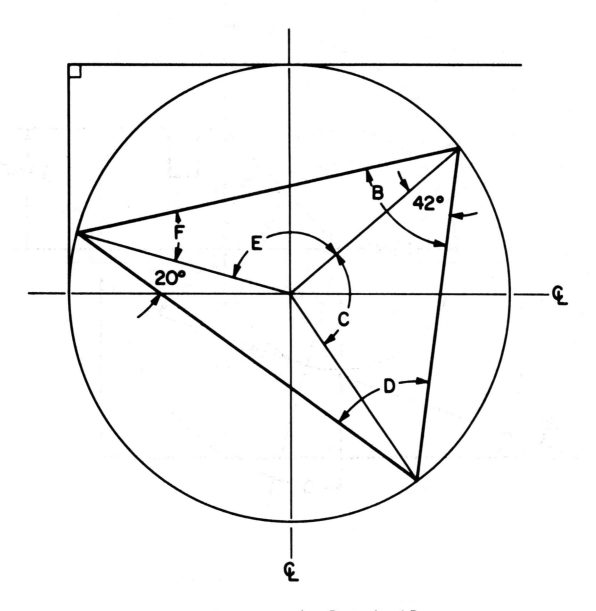

1. Determine ∠D = _____
2. Determine ∠E = _____
3. Determine ∠F = _____
4. Determine ∠B = _____
5. Determine ∠C = _____

Problem 9

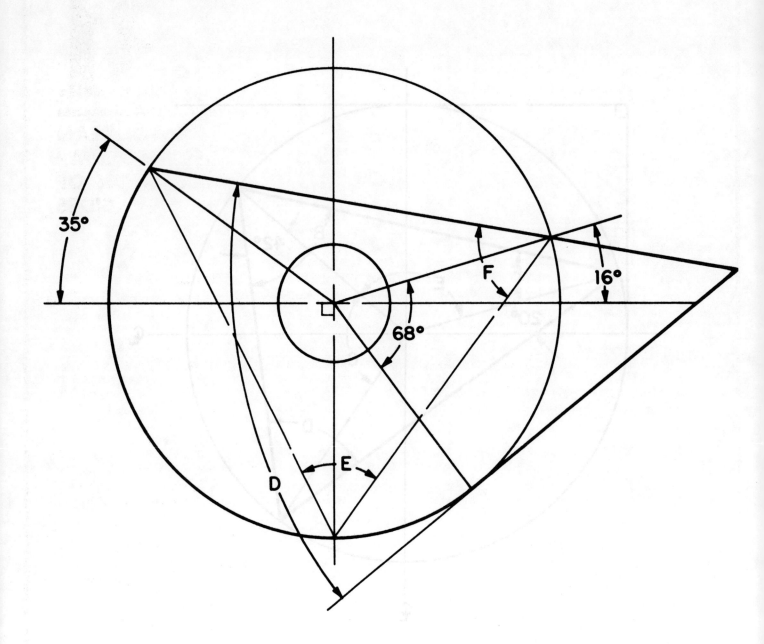

1. Determine ∠D = _____
2. Determine ∠E = _____
3. Determine ∠F = _____

UNIT 31

PYTHAGOREAN THEORUM - PROJECTION OF SIDES

INTRODUCTION

The first task a machinist performs after receiving a blueprint is to analyze various ways the job can be done. A skilled machinist seeks out the lines and angles to be calculated which often are in the form of triangles. This unit explains two methods for solving triangles to determine the angles and length of sides. The first is called the Pythagorean Theorem which applies for right triangles. The second is the projection of sides process which will solve any triangle. A good understanding of both these principles is essential to the machinist.

OBJECTIVES:

After completing this unit the student will be able to:

- Determine the length of any side of a right triangle using the Pythagorean Theorem
- Apply the **projection of sides proposition**

PYTHAGOREAN THEOREM

PROPOSITION XIII:

The square of the hypotenuse of a right triangle is equal to the sum of the squares of the legs.

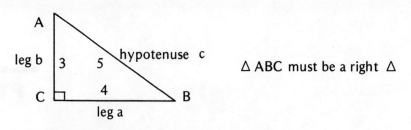

△ ABC must be a right △

FORMULA: $a^2 + b^2 = c^2$

GIVEN:

$a = 4$, $b = 3$ ⟶ find $c = $ ___?___

If			then	
$a^2 + b^2$	=	c^2		Given
$4^2 + 3^2$	=	c^2		substitution
$16 + 9$	=	c^2		substitution
25	=	c^2		substitution
$\sqrt{25}$	=	$\sqrt{c^2}$	∴	$\sqrt{}$ axiom of equality
5	=	c		substitution

Similar rules may be used to determine the length of a or b :

1. $a^2 = c^2 - b^2$
2. $b^2 = c^2 - a^2$

216

PROPOSITION XIV:

The projection of a side of a triangle upon the base is equal to the square of this side plus the square of the base minus the square of the third side, all divided by two times the base.

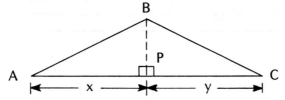

(B projects onto point P and x is the projection of AB on AC)

FORMULA: $$x = \frac{(AB)^2 + (AC)^2 - (BC)^2}{2(AC)}$$

FORMULA: $$y = \frac{(BC)^2 + (AC)^2 - (AB)^2}{2(AC)}$$

EXAMPLE 1:

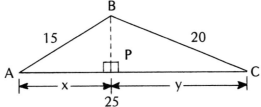

Find x = ?

$$x = \frac{15^2 + 25^2 - 20^2}{2 \cdot 25} =$$

$$x = \frac{225 + 625 - 400}{50}$$

$$x = \frac{850 - 400}{50}$$

$$x = \frac{450}{50}$$

$$x = 9$$

EXAMPLE 2:

Find y = ?

$$y = \frac{20^2 + 25^2 - 15^2}{2 \cdot 25}$$

$$y = \frac{400 + 625 - 225}{50}$$

$$y = \frac{1025 - 225}{50}$$

$$y = \frac{800}{50}$$

$$y = 16$$

SAMPLE PROBLEM

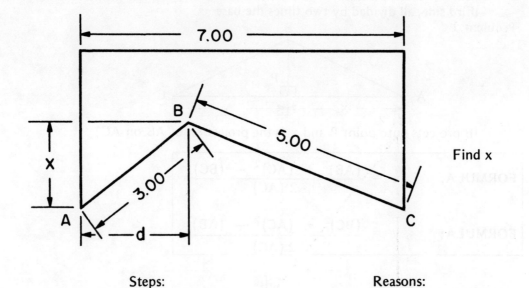

Find x

Steps:	Reasons:
1) First, find the projection of AB along the horizontal. $d = \dfrac{\overline{AB}^2 + \overline{AC}^2 - \overline{BC}^2}{2\overline{AC}}$ $= \dfrac{9 + 49 - 25}{14}$ $= \dfrac{33}{14}$, or 2.36	1) In a triangle, the projection of a side upon the base is equal to the square of this side plus the square of the base minus the square of the third side divided by twice the base.
2) Then, in the right $\triangle$ formed by AB, d and x; $AB = d^2 + x^2$ (3) $= \left(\dfrac{33}{14}\right)^2 + x$ $x = 3.44 +$ $x = \sqrt{3.44} +$ $x = 1.85 +$	2) The square of the hypotenuse is equal to the sum of the squares of the two legs..

CONCEPT APPLICATIONS:

Problem 1

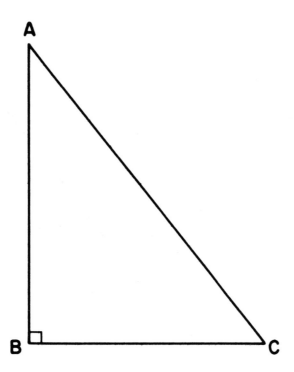

1. Find the length of AB if the length of AC = 3.700 and the length of BC = 2.400. AB = _____ .

2. Assume the length of each leg of the right triangle is equal to 2.800. Calculate the length of the hypotenuse. AC = _____ .

3. Assume the length of the hypotenuse is equal to 5.750. What would be the length of each of the legs if they were equal to each other?
AB = BC = _____ .

4. Assume the length of leg BC is $1\frac{1}{2}$ times that of leg AB and the hypotenuse is equal to 6.500. Calculate the length of each leg.

Problem 2

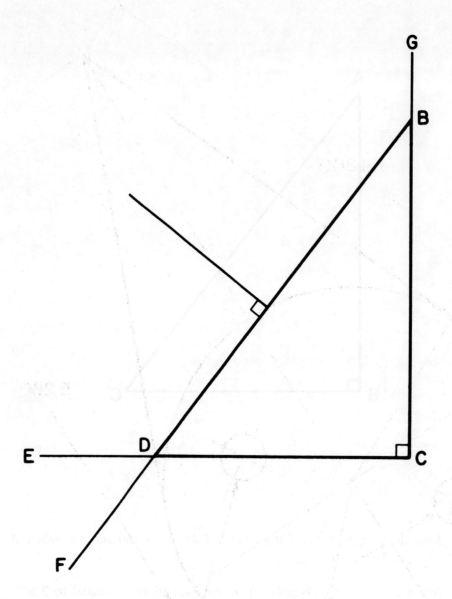

1. Determine ∠BDC in right △BCD if ∠DBC = 38°. _____
2. Determine ∠DBE = _____
3. Determine ∠EDF = _____
4. Determine ∠DBG = _____

Problem 3

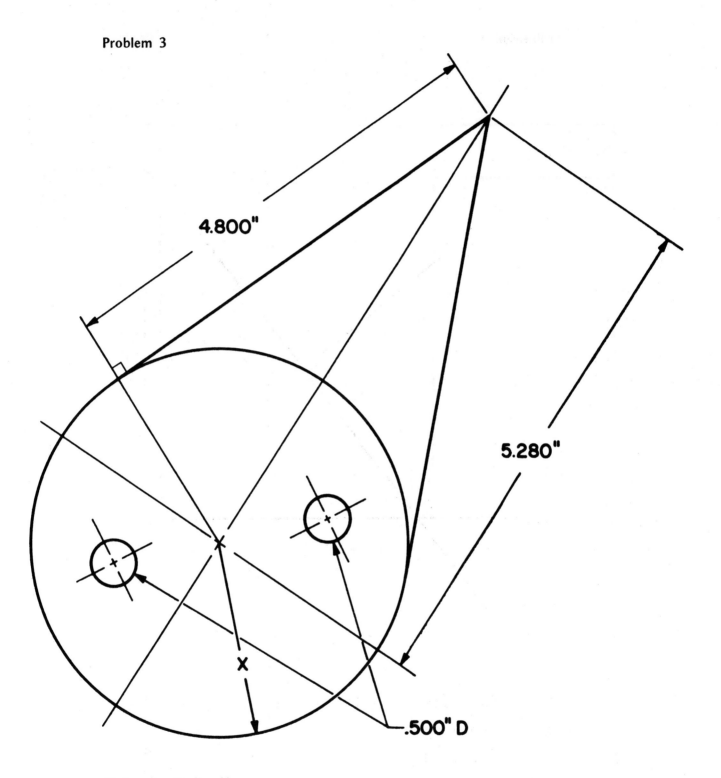

Determine Radius X = _____

Problem 4

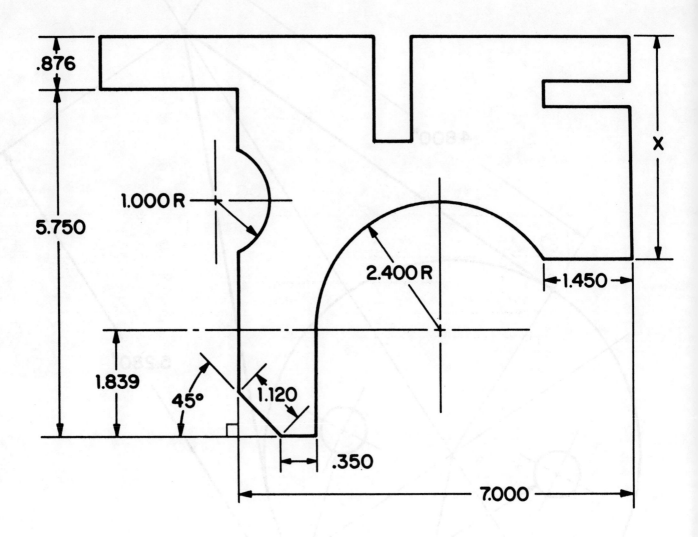

Use the 45° angle as the hypotenuse. The 2.400 can be used as the hypotenuse of another right triangle.

Calculate the distance X = _____

Problem 5

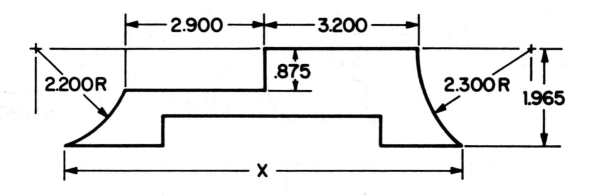

Determine Distance X = _____

Problem 6

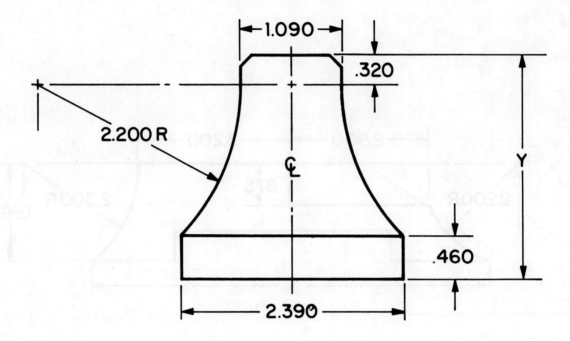

The figure is symmetrical about its vertical center line.

Hint: 2.200R becomes the hypotenuse.

Calculate distance Y = _____

Problem 7

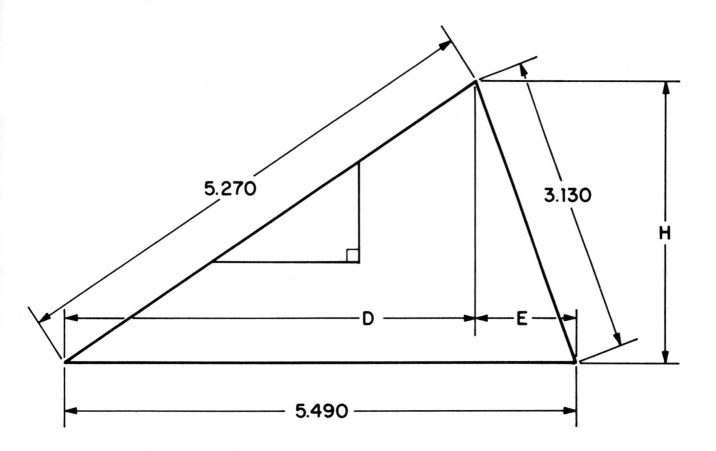

1. Calculate the distance D = _____

    Hint: Use the projection formula.

2. Calculate the distance E = _____

3. Calculate the altitude H = _____

Problem 8

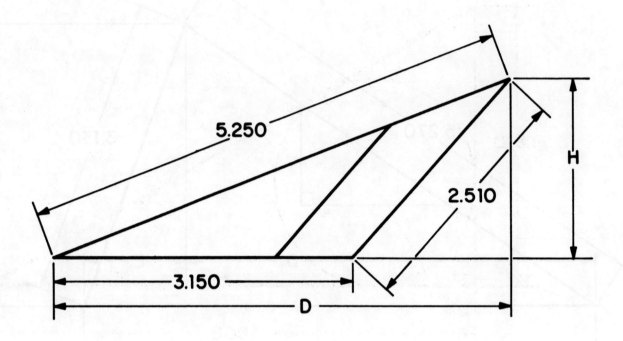

Calculate the distance D = _____

Calculate the altitude H = _____

226

Problem 9

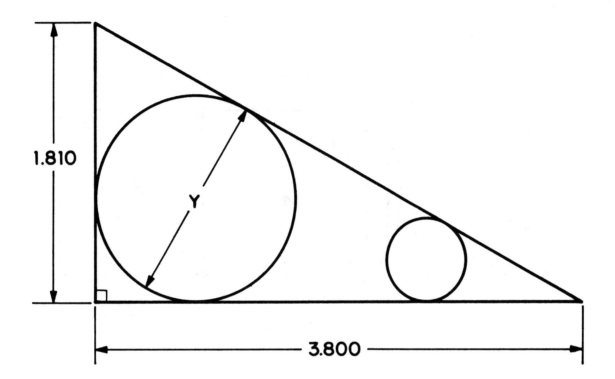

Solve for Y = _____

227

Problem 10

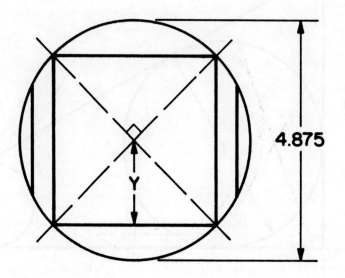

Solve for Y = _____

Problem 11

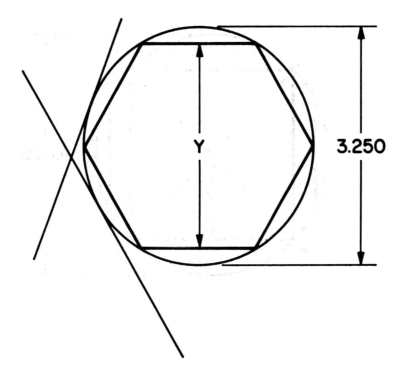

Solve for Y = _____

(That is, find distance across the flats of the inscribed hexagon.)

229

Problem 12

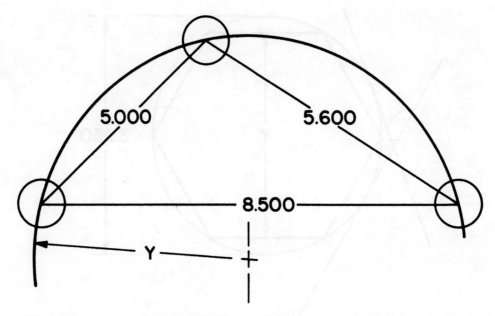

NOTE: When a triangle is inscribed in a circle the diameter equals the product of the sides (chords) divided by the altitude of the triangle.

Solve for Y = _____

Problem 13

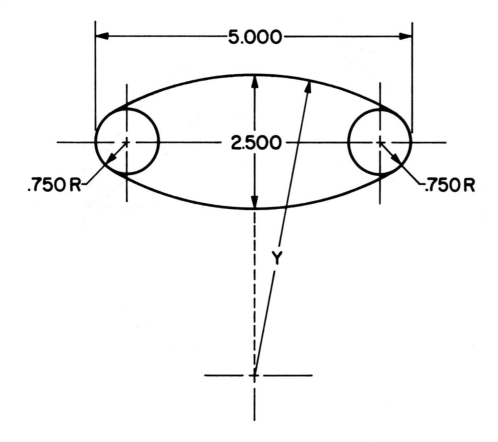

Solve for Y = _____

UNIT 32
CIRCLES

INTRODUCTION

The machinist works daily with radii (plural of radius), diameters, circimferences and angular radial dimensions. Selecting milling cutter sizes, calculating speeds and making measurements between centers or across threads are all operations which involve circles. Thus, a thorough understanding of the circle and its parts is fundamental to the geometry of machining.

OBJECTIVES:

After completing this unit the student will be able to:

- Identify the basic parts of a circle
- Define basic terms used in circular measurement

DEFINITION:

 CIRCLE

The set of all points which are equidistant from a fixed point in a plane.

(fixed point A is the origin of circle A)

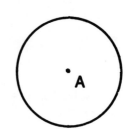

✋ **RADIUS**

A segment which joins the origin to any point on the circle. ($\overline{AB}$)

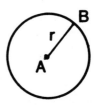

🛑 DIAMETER

A segment which joins 2 points of the circle and contains (passes through) the origin. ($\overline{BC}$) (diameter = 2 radii)

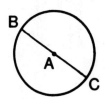

🛑 CHORD

A segment which joins any 2 points of the circle. ($\overline{AB}$)

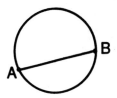

🛑 CIRCUMFERENCE

The distance around the circle. (C)

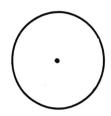

🛑 ARC

A curved segment of the circle: $\overset{\frown}{AB}$

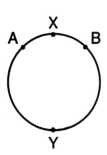

NOTE:

An **arc** is frequently named using three letters: $\overset{\frown}{AXB}$ or $\overset{\frown}{AYB}$ to eliminate confusion.

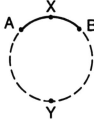

🛑 CENTRAL ANGLE

An angle formed by two radii using the origin as the vertex of the angle: ($\angle ABC$)

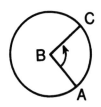

✋ INSCRIBED ANGLE

An angle formed by two chords using any point on the circle as the vertex of the angle: ($\angle ABC$)

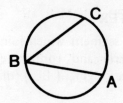

✋ SEGMENT OF A CIRCLE

That part of the circle bounded by an arc and its chord: (shaded region)

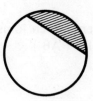

✋ SECTOR OF A CIRCLE

That part of the circle bounded by the radii and an arc: (shaded region)

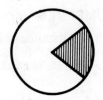

✋ TANGENT

A line which intersects the circle in only one point: (P) (Read: ℓ is tangent to circle O at point P)

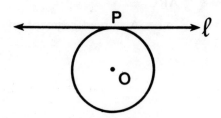

✋ SECANT

A line which intersects the circle in exactly two points: (ℓ and m are both secants)

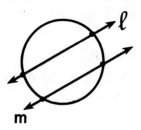

RELATED FORMULAS (Circumference and Area)

💡 Circumference (C) of a circle is equal to the product of pi (π) and the diameter (d).

$$(C = \pi d)$$
$$(\text{Use } \pi = 3\tfrac{1}{7} = \tfrac{22}{7} \text{ or } 3.1416)$$

EXAMPLE:

If d = 21", find C = ____?

$C = \pi d$

$C = \dfrac{22}{7} \cdot \dfrac{21}{1}$

$C = 22 \cdot 3$

$C = 66"$

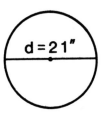

💡 Area of a circle is equal to pi (π) times the square of the radius (r^2).

$$(A = \pi r^2)$$

EXAMPLE:

$A = \pi r^2$

$A = \dfrac{22}{7} \cdot \dfrac{7^2}{1}$

$A = \dfrac{22}{\cancel{7}} \times \cancel{49}^{7}$

$A = 154$ square inches

💡 **PROPOSITION XV:**

A straight line perpendicular to a radius at its end point is tangent to the circle.

OBSERVATIONS:

1. $\ell \perp \overline{OP}$
2. $\angle 1 = \angle 2 = 90°$
3. P is the point of tangency

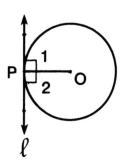

PROPOSITION XVI:

If two circles are tangent internally, the segment joining their center points is equal to the difference of their radii and if extended will pass through the point of tangency.

or

$(\overline{PO} = R - r)$

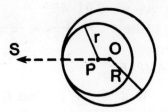

PROPOSITION XVII:

If two circles are tangent externally, the segment joining their center points is equal to the sum of their radii and passes through the common point of tangency.

or

$(\overline{PO} = R + r)$

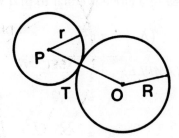

PROPOSITION XVIII:

A line segment drawn from the center of the circle perpendicular to a chord bisects the chord.

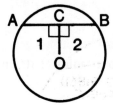

$\overline{OC} \perp \overline{AB}$
$\overline{AC} = \overline{BC}$
$\angle 1 = \angle 2$

PROPOSITION XVIX:

The central angle has the same measure as its intercepted arc (in degrees).

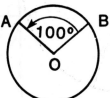

$\angle AOB$ is the central $\angle$.
$\angle AOB = \overparen{AB}$ in degrees

PROPOSITION XX:

An inscribed angle is equal to one half its intercepted arc.

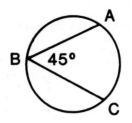

∠ ABC is the inscribed ∠.

∠ ABC = $\frac{1}{2}$ $\overarc{AC}$ in degrees

SUMMARY OF GEOMETRY

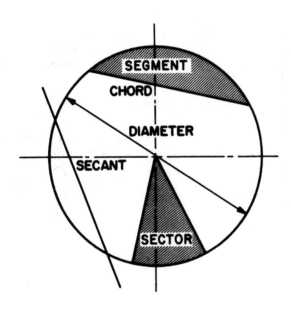

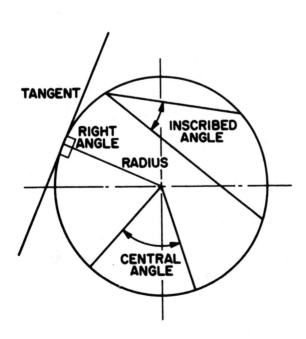

SUMMARY OF BASIC AXIOMS OF GEOMETRY

1. Things equal to the same thing are equal to each other.

2. Any number or quantity may be substituted for its equal.

3. If equals are added to equals, the results are equals.

4. If equals are subtracted from equals, the results are equals.

5. If equals are multiplied by equals, the results are equal.

6. If equals are divided by equals, the results are equals.

7. A whole unit is greater than any of its parts and is equal to the sum of its parts.

8. Two points fix the position of a line. Only one line can be drawn through 2 given points.

9. Two straight lines in the same plane can only intersect in 1 point: (at most).

10. Given a line, and a point which is not on that line, there exists exactly one other line, passing though that point which is also parallel to the given line.

SUMMARY OF PROPOSITIONS OF GEOMETRY

PROPOSITION I: If two straight lines intersect, the opposite or vertical angles are equal.

PROPOSITION II: If two or more lines in the same plane are perpendicular to the same line, they are parallel to each other.

PROPOSITION III: If two parallel lines are cut by a third line (a transversal), then the pairs of alternate interior angles are equal.

PROPOSITION IV: If two parallel lines are cut by a third line (a transversal), then the pairs of alternate exterior angles are equal.

PROPOSITION V: If two parallel lines are cut by a third line (a transversal), then the pairs of corresponding angles are equal.

PROPOSITION VI: The sum of the interior angles of a triangle is 180°

SUMMARY OF PROPOSITIONS OF GEOMETRY (CONTINUED)

PROPOSITION VII: An exterior angle of a triangle is equal to the sum of the two interior angles at the other vertices.

PROPOSITION VIII: If two angles have their corresponding sides parallel, then the angles are equal.

PROPOSITION IX: If two angles have their corresponding sides perpendicular, then the angles are equal.

PROPOSITION X: The sum of the interior angles of a polygon is equal to 180° times the number of sides less 2 [or $S = 180°(n - 2)$] .

PROPOSITION XI: A line segment drawn from a point of tangency to the center of a circle is perpendicular to the tangent.

PROPOSITION XII: If two tangents are drawn to a circle from the same exterior point, then the corresponding segments are equal and a line from this external point to the center of the circle forms two equal angles.

PROPOSITION XIII: The square of the hypotenuse of a right triangle is equal to the sum of the squares of the legs.

PROPOSITION XIV: The projection of a side of a triangle upon the base is equal to the square of this side plus the square of the base minus the square of the third side, all divided by two times the base.

PROPOSITION XV: A straight line perpendicular to a radius at its end point is tangent to the circle.

PROPOSITION XVI: If two circles are tangent internally, the segment joining their center points is equal to the difference of their radii and if extended will pass through the point of tangency.

PROPOSITION XVII: If two circles are tangent externally, the segment joining their center points is equal to the sum of their radii and passes through the common point of tangency.

PROPOSITION XVIII: A line segment drawn from the center of the circle perpendicular to a chord bisects the chord.

PROPOSITION XVIX: The central angle has the same measure as its intercepted arc (in degrees).

PROPOSITION XX: An inscribed angle is equal to one half its intercepted arc.

CONCEPT APPLICATIONS:

Problem 1

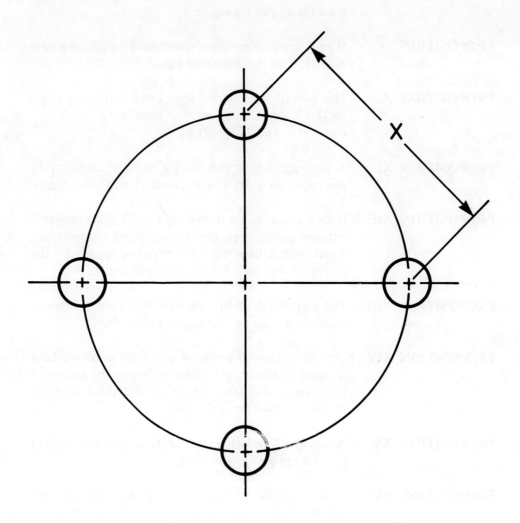

Four equally spaced holes on a 3.500 diameter bolt circle.

Find Distance X = _____

Problem 2

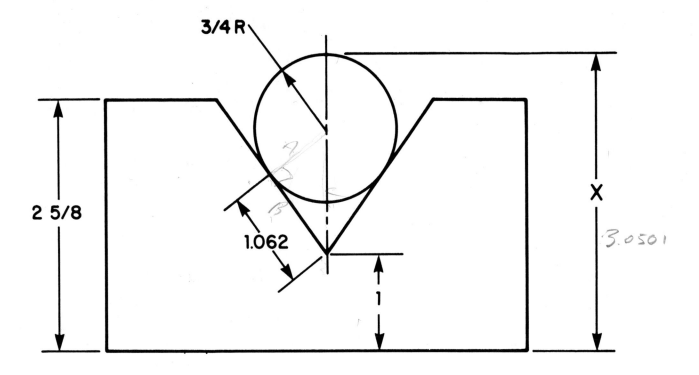

Find Distance X = _____

Problem 3

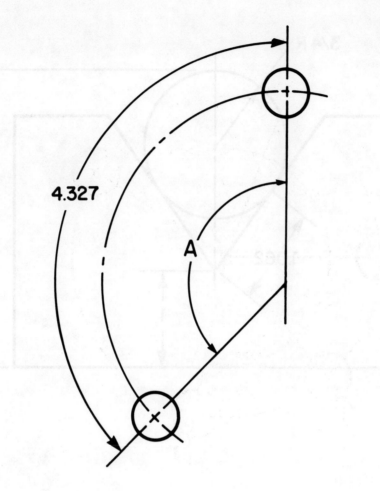

Two holes are spaced 4.327 apart on a bolt circle that is 3.875 in diameter.

What is angle A = _____

Problem 4

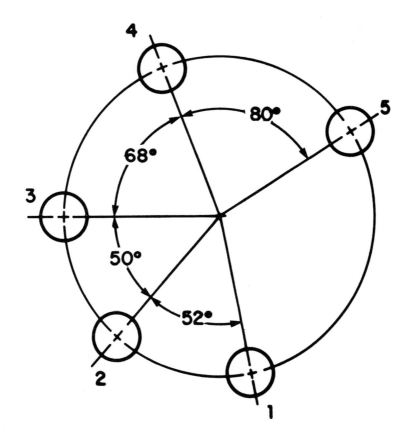

It is required to drill 5 holes on a 3.312 diameter bolt circle. To maintain the angles specified determine the arc distance through which the rotary table be rotated for each hole.

Arc 1–2 = _____

Arc 2–3 = _____

Arc 3–4 = _____

Arc 4–5 = _____

Arc 5–1 = _____

Problem 5

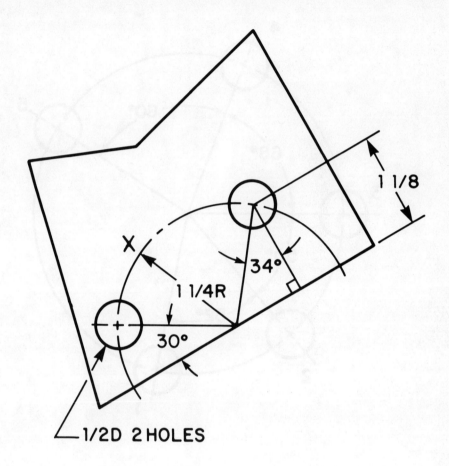

What is the length of arc between centers of the two holes?

X = _____

TRIGONOMETRY

$$\frac{b}{c} = \frac{adj}{hyp}$$

UNIT 33

INTRODUCTION TO TRIGONOMETRY

INTRODUCTION

Trigonometry is used to solve triangles when specific relationships are known about the angles or sides. Many practical problems that cannot be solved by geometry alone can easily be solved with trigonometry. The basis of trigonometry relates to an angle and a ratio of two sides of the triangle. Developing strong skills in trigonometry is an important complement to the skills you have learned in geometry. Many problems entail using combinations of both skills. This is the test of the true skilled craftsman in the machining industry.

OBJECTIVES:

After completing this unit the student will be able to:

- Name and label the basic parts of a right triangle

Trigonometry comes from the Greek words: tri — (three), gonia — (angle), and metron — (measure); that is, "triangle measure". The triangle used here is the right triangle. Each part of the triangle has a special name:

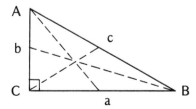

- The capital letters name the vertex points with "C" always assigned to the right angle.

- Lower case letters are assigned as measures of the three sides. They must be located opposite the corresponding capital letters. (See the diagram above).

247

DEFINITIONS: (for rt. △, sides of)

🛑 OPPOSITE SIDE (opp.)

The side which is opposite the given angle:

Side a is opposite (opp.) angle A.

Side b is opposite angle B.
(Sides a and b are called legs)

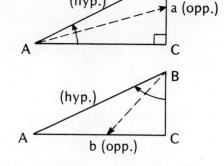

🛑 ADJACENT SIDE (adj.)

The side which forms the acute angle with the hypotonuse:

Side b is adjacent (adj.) to angle A.

Side a is adjacent to angle B.

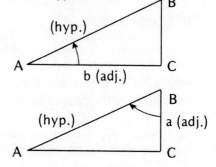

🛑 HYPOTENUSE

The side opposite the right angle. The hypotenuse is always the longest side of the triangle (name with the small letter c).

💡 The sides (legs) a and b may be either opposite or adjacent. The position of the side in relation to the angle determines which it is:

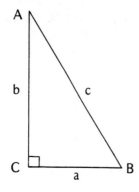

In △ ABC
side a is opp. ∠ A
side a is adj. to ∠ B
side b is opp. ∠ B
side b is adj. to ∠ A
side c is the hypotenuse
side c is opp. the rt. ∠ C

CONCEPT APPLICATIONS:

Label the triangles according to standard practice.

(1)
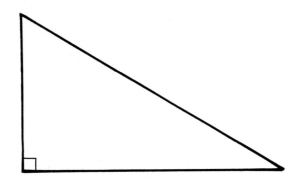

(2)
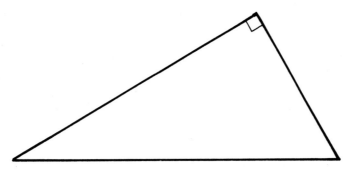

(3) 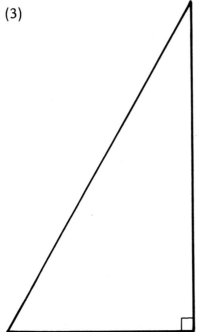 (4)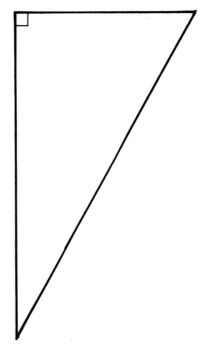

UNIT 34

FUNCTIONS OF ANGLES

INTRODUCTION

The trigonometric function of an angle is an expression of the ratio between any two sides of a triangle. Each angle has six ratios and six functions that define the size of the angle. An understanding of the six basic formulas or functions will enable you to solve any triangle if either an angle and a side or a ratio of the sides is known.

OBJECTIVES:

After completing this unit the student will be able to:

- Construct ratios of sides for a given angle of a right triangle
- Name the functions of a given angle using the ratios of the sides of a right triangle

RATIO OF SIDES OF AN ANGLE

A ratio is a fraction used to relate the measures of any 2 sides of a right triangle.

EXAMPLE 1:

Determine the six possible ratios in rt. △ABC.

Given: △ABC with sides of:

a = 3"
b = 4"
c = 5"

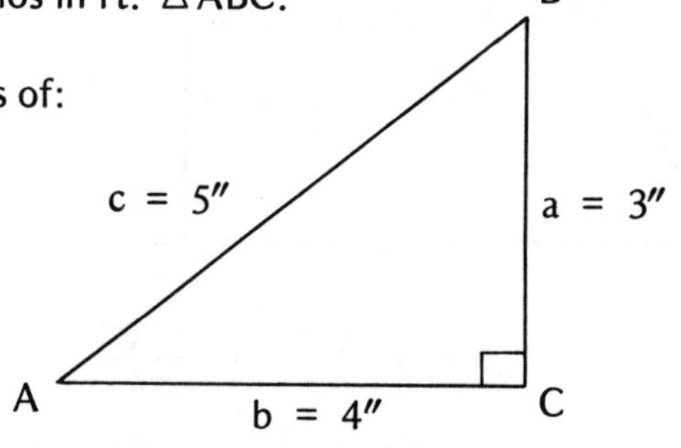

The six possible ratios would be:

$$\left\{\frac{a}{c}, \frac{b}{c}, \frac{a}{b}, \frac{b}{a}, \frac{c}{b}, \frac{c}{a}\right\}$$

or expressed numerically:

$$\left\{\frac{3}{5}, \frac{4}{5}, \frac{3}{4}, \frac{4}{3}, \frac{5}{4}, \frac{5}{3}\right\}$$

EXAMPLE 2:

Angle A is an **acute angle** and is determined or measured as follows:

CONSIDER:

The angle between zero and 90° such that AC = 6, BC = 4.

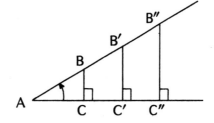

The relationship between the measures of these sides can be written as a fraction:

$$\frac{BC}{AC} = \frac{4}{6} = \frac{2}{3} = \frac{a}{b}$$

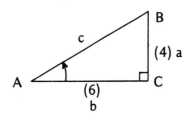

Extend the sides: Extend again:

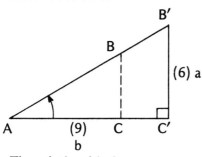 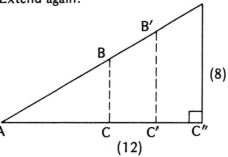

The relationship is now:

$$\frac{B'C'}{AC'} = \frac{6}{9} = \frac{2}{3} = \frac{a}{b} \qquad \frac{B''C''}{AC''} = \frac{8}{12} = \frac{2}{3} = \frac{a}{b}$$

NOTE:

$$\frac{BC}{AC} = \frac{B'C'}{AC'} = \frac{B''C''}{AC''} = \frac{2}{3} \text{ in each case,}$$

the ratio did not change and ∠A did not change in size.

A basic function, such as in this example, is one wherein an angle measure remains constant and the ratio of the lengths of corresponding parts of similar right triangles also remains constant.

FUNCTIONS OF AN ANGLE

There are 6 basic ratios or functions in every right triangle as follows:

$$\frac{a}{c}, \frac{b}{c}, \frac{a}{b}, \frac{b}{a}, \frac{c}{b}, \frac{c}{a}.$$

and these functions are called:

$\frac{a}{c}$ = sine of ∠A, written as: $\sin \angle A = \frac{opp.}{hyp.}$

$\frac{b}{c}$ = cosine of ∠A, written as: $\cos \angle A = \frac{adj.}{hyp.}$

$\frac{a}{b}$ = tangent of ∠A, written as: $\tan \angle A = \frac{opp.}{adj.}$

$\frac{b}{a}$ = cotangent of ∠A, written as: $\cot \angle A = \frac{adj.}{opp.}$

$\frac{c}{b}$ = secant of ∠A, written as: $\sec \angle A = \frac{hyp.}{adj.}$

$\frac{c}{a}$ = cosecant of ∠A, written as: $\csc \angle A = \frac{hyp.}{opp.}$

EXAMPLE 1: Find the values of the functions of angle A

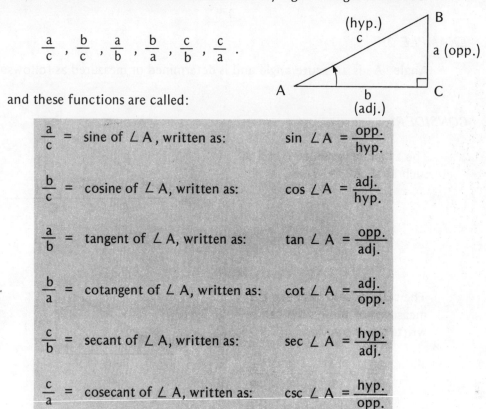

$\frac{a}{c} = \frac{3}{5} = .600 = \sin \angle A$

$\frac{b}{c} = \frac{4}{5} = .800 = \cos \angle A$

$\frac{a}{b} = \frac{3}{4} = .750 = \tan \angle A$

$\frac{b}{a} = \frac{4}{3} = 1.333 = \cot \angle A$

$\frac{c}{b} = \frac{5}{4} = 1.250 = \sec \angle A$

$\frac{c}{a} = \frac{5}{3} = 1.666 = \csc \angle A$

EXAMPLE 2:   Find the set of ratios for angle B

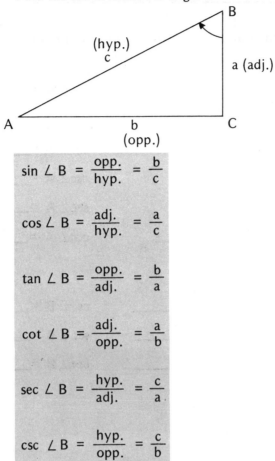

$$\sin \angle B = \frac{opp.}{hyp.} = \frac{b}{c}$$

$$\cos \angle B = \frac{adj.}{hyp.} = \frac{a}{c}$$

$$\tan \angle B = \frac{opp.}{adj.} = \frac{b}{a}$$

$$\cot \angle B = \frac{adj.}{opp.} = \frac{a}{b}$$

$$\sec \angle B = \frac{hyp.}{adj.} = \frac{c}{a}$$

$$\csc \angle B = \frac{hyp.}{opp.} = \frac{c}{b}$$

NOTE:

The opposite side of ∠B is side b and the opposite side of ∠A is side a. The formula for the sine of any angle (θ) is: $\sin \theta = \frac{opp.}{hyp.}$. A common source of error in computing the function of an angle is write "opp" and "adj" on your worksheet and then use these measures incorrectly in the formula. In finding the functions above for ∠A and ∠B, the sine function of sin ∠A ≠ sin ∠B, since $\sin \angle A = \frac{a}{c}$ and $\sin \angle B = \frac{b}{c}$. Be observant and avoid this pitfall. Hint: Memorize functions in terms of adjacent, opposite and hypotenuse but do not write those terms on your worksheet triangle.

CONCEPT APPLICATIONS:

Indicate the six functions of each acute angle in ratio form:

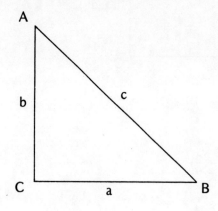

sin ∠A = _____
cos ∠A = _____
tan ∠A = _____
csc ∠A = _____
sec ∠A = _____
cot ∠A = _____

sin ∠B = _____   csc ∠B = _____
cos ∠B = _____   sec ∠B = _____
tan ∠B = _____   cot ∠B = _____

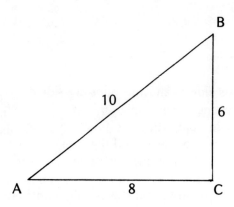

sin ∠A = _____
cos ∠A = _____
tan ∠A = _____
csc ∠A = _____
sec ∠A = _____
cot ∠A = _____

sin ∠B = _____   csc ∠B = _____
cos ∠B = _____   sec ∠B = _____
tan ∠B = _____   cot ∠B = _____

THE UNIT CIRCLE

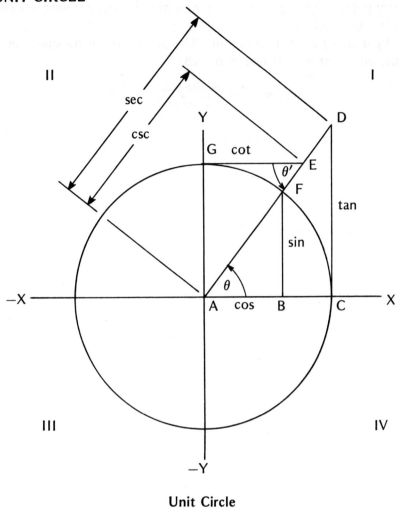

Unit Circle

The unit circle is used to explain how the six trigonometric functions (sine, cosine, tangent, cotangent, secant and cosecant) are derived. Referring to the unit circle illustration, you will notice that the circle is divided into four equal sectors of 90°. These sectors, commonly referred to as quadrants, are numbered I, II, III and IV counterclockwise, starting with the upper right hand quadrant first. The importance of those four quadrants is due mainly to the positive and negative nature of a particular trigonometric function through 360°. Before we proceed on to the theory of how a trigonometric function changes its value, explanation of each function is necessary.

In the figure we have three distinct triangles of reference; they are △FAB, △DAC, and △GAE. ∠θ = ∠θ', lines AG, AC and AF have a measure of one (1). Line AF is positive (+) throughout 360°. Using △FAB, we will begin by deriving the sine function. The definition of the **sine** is the opposite divided by the hypotenuse in △FAB:

$$\sin \angle \theta = \frac{\text{opposite side}}{\text{hypotenuse}} = \frac{FB}{AF}$$

Since AF is equal to one (1), we can therefore substitute (1) for AF and solve the equation:

$$\sin \angle \theta = \frac{FB}{1} = FB$$

∴ **SINE** is equal to line FB in △FAB.

The cosine function is also derived by using △FAB. By definition, the **COSINE** is equal to:

$$\cos \angle \theta = \frac{\text{adjacent side}}{\text{hypotenuse}} = \frac{AB}{AF}$$

Since AF is equal to one (1), we can substitute one (1) for AF in the equation.

$$\text{cosine} \angle \theta = \frac{AB}{1} = AB$$

∴ **COSINE** is equal to line AB in △FAB.

In deriving the tangent and secant functions, we will be using △DAC as our reference. By definition, tangent is equal to the opposite side divided by the adjacent side of △DAC.

$$\tan \angle DAC = \frac{\text{opposite side}}{\text{adjacent side}} = \frac{DC}{AC}$$

Since AC is equal to one (1), we can substitute one (1) in our equation.

$$\tan \angle \theta = \frac{DC}{1} = DC$$

∴ **TANGENT** is equal to line DC in △DAC.

✋ The **SECANT**, by definition is the hypotenuse divided by the adjacent side in △DAC.

$$\sec \angle \theta = \frac{\text{hypotenuse}}{\text{adjacent}} = \frac{AD}{AC}$$

Since AC is equal to one (1), we can therefore substitute one (1) in our equation.

$$\sec \angle \theta = \frac{AD}{1} = AD$$

∴ **SECANT** equal line AD in △DAC.

✋ The cotangent and cosecant are derived by using △GAE. By definition, the **COTANGENT** is equal to the adjacent side divided by the opposite side in △GAE.

$$\cot \angle \theta' = \frac{\text{adjacent side}}{\text{opposite side}} = \frac{GE}{AG}$$

Since AG is equal to one (1), we can substitute one (1) in our equation.

$$\cot \angle \theta' = \frac{GE}{1} = GE$$

∴ **COTANGENT** is equal to line GE in △GAE.

✋ By definition, **COSECANT** is equal to the hypotenuse divided by the opposite side in △GAE.

$$\csc \angle \theta' = \frac{\text{hypotenuse}}{\text{opposite side}} = \frac{AE}{AG}$$

Since AG is equal to one (1), we can therefore substitute one (1) in our equation.

$$\csc \angle \theta' = \frac{AE}{1} = AE$$

∴ **COSECANT** is equal to line AE in △GAE.

The same method for deriving these six trigonometric functions is used for quadrants II, III, and IV.

POSITIVE AND NEGATIVE VALUES OF THE SIX TRIGONOMETRIC FUNCTIONS IN QUADRANTS I THROUGH IV

As should be noted when deriving the trigonometric functions in quadrant I, all the values are positive. This is because our y and x segments of the co-ordinate system are positive with respect to their location.

In quandrant II, y will remain positive (+) and x changes to negative (−), in quadrant III both x and y are negative, in quadrant IV, y remains negative (−) and x changes to positive (+). (You may want to review Unit 17, signed numbers).

Referencing the derivation of the six trigonometric functions in quadrant I, it should be noted that all of the functions were positive in nature. However, this would not be the case for the remaining three quadrants. The changing of a sign from positive to negative and vice-versa for a trigonometric function is a result of the Cartesian coordinate system.

In this system, there are two lines, the x and y axis, that meet perpendicular to each other at a point called the origin (0) and this point divides the two axes into positive and negative segments. This point also establishes our four quadrants of reference, 90° each, and is used to determine the positive and negative value of a trigonometric function.

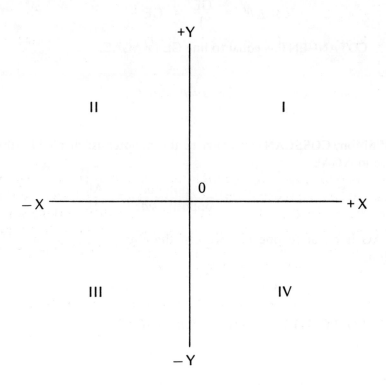

Cartesian Coordinate System

Only the four quadrants are shown in this illustration with the unit circle and geometric construction eliminated. The vertical line (the y axis) has a positive (+) leg and a negative (−) leg; both originating at the zero (0) point and identified by +y and −y respectively. The horizontal line (the x axis) also has a positive (+) and a negative (−) leg originating at the zero point and is identified by +x and −x respectively. Knowing each axis leg value, we can determine the value of each trigonometric function in the four quadrants. Since it is not necessary to derive all of the trigonometric functions for quadrant II, III, and IV, a simple reference chart (Table 1) has been designed for the purpose of readily determining the positive or negative value of a trigonometric value in a particular quadrant.

	I	II	III	IV
	0°– 90°	90°00'01"– 180°	180°00'01"– 270°	270°00'01"– 360°(0)
SIN	Pos (+)	Pos (+)	Neg (−)	Neg (−)
COS	Pos (+)	Neg (−)	Neg (−)	Pos (+)
TAN	Pos (+)	Neg (−)	Pos (+)	Neg (−)
COT	Pos (+)	Neg (−)	Pos (+)	Neg (−)
SEC	Pos (+)	Neg (−)	Neg (−)	Pos (+)
CSC	Pos (+)	Pos (+)	Neg (−)	Neg (−)

Table 1: Sign of functions in Cartesian system

RELATED ANGLES

At one time or another, angles greater than 90° will be encountered. In solving trigonometric equations, usually most text book tables list natural trigonometric functions from 0° to 180°. However when working with angles greater than 180°, a certain degree of mathematical manipulation is required. Usually this consists of reducing an angle greater than 90° to its image or related angle, or an angle between 0° and 90°.

To accomplish this, a simple method of calculating the RELATED ANGLE has been developed and is shown in Table 2.

90° − 180°	180° − 270°	270° − 360°
180° − ∠X	∠X − 180°	360° − ∠X

Table 2: Related Angle

EXAMPLE:

What is the related angle for 240°? Referring to Table 2, we would substract 180° from 240° leaving us;

$$\angle X - 180° =$$

$$240° - 180° = 60°$$

with a related angle of 60°. In order to determine sin 240° we would then use the sin 60° because it is the image or related angle of 240° (figure below) and its SIGN would be negative (−) because it is a third quadrant angle (Table 1).

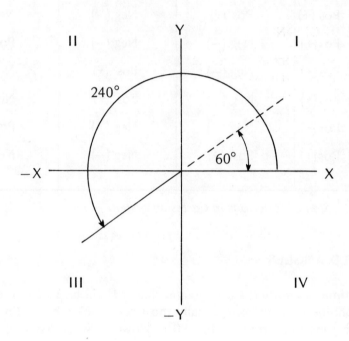

UNIT 35

VALUES OF FUNCTIONS

INTRODUCTION

Each basic trigonometric function for one angle of a right triangle can be expressed as equal to a similar function of the complementary angle. The values of these functions are found in tables called "Table of Natural Trigonometric Functions" and can also be determined on most hand held calculators available today. Use of the calculator is rapidly replacing the use of the trigonometric tables. However, you should be familiar with both methods.

OBJECTIVES:

After completing this unit the student will be able to:

- Express each basic function as equal to another function in the same triangle
- Determine the value of any given function

Equal functions have the same numerical value. Each basic function for Angle A can be expressed as equal to some other function of angle B. There are six of these pairs in Right Triangle ABC:

1. $\sin \angle A = \cos \angle B = \dfrac{a}{c}$
2. $\cos \angle A = \sin \angle B = \dfrac{b}{c}$
3. $\tan \angle A = \cot \angle B = \dfrac{a}{b}$
4. $\cot \angle A = \tan \angle B = \dfrac{b}{a}$
5. $\sec \angle A = \csc \angle B = \dfrac{c}{b}$
6. $\csc \angle A = \sec \angle B = \dfrac{c}{a}$

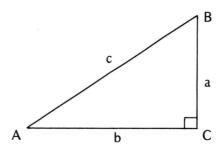

EXAMPLE 1:

$$\tan \angle A = \frac{opp}{adj} = \frac{a}{b}$$

$$\cot \angle A = \frac{adj}{opp} = \frac{a}{b}$$

THEREFORE:

$$\tan \angle A = \cot \angle B$$

TO FIND THE VALUE OF A FUNCTION:

Step 1 Identify the sides of the triangle to be used in the function.

Step 2 Construct the ratio of the lengths of these sides.

Step 3 Change this ratio into decimal form (by dividing the denominator into the numerator and add zeros as needed.)

EXAMPLE 2:

In the right triangle ABC:

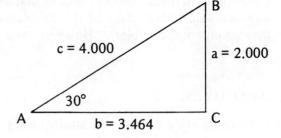

$$\sin \angle A = \sin 30° = \frac{a}{c} = \frac{2}{4} = .500$$

$$\cos \angle A = \cos 30° = \frac{b}{c} = \frac{3.464}{4} = .866$$

$$\tan \angle A = \tan 30° = \frac{a}{b} = \frac{2}{3.464} = .577$$

$$\cot \angle A = \cot 30° = \frac{b}{a} = \frac{3.464}{2} = 1.732$$

$$\sec \angle A = \sec 30° = \frac{c}{b} = \frac{4}{3.464} = 1.154$$

$$\csc \angle A = \csc 30° = \frac{c}{a} = \frac{4}{2} = 2.000$$

CONCEPT APPLICATION:

Complete the following exercise and compare the answers with those of Example 2.

NOTE:

Angle A and angle B are complementary angles (the sum is 90°).

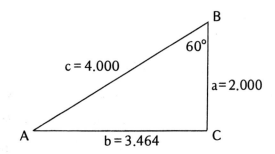

$$\sin \angle B = \sin 60° = \frac{b}{c} = \frac{3.464}{4} = .866$$

$$\cos \angle B = \cos 60° = \underline{} = \underline{} = \underline{}$$

$$\tan \angle B = \tan 60° = \underline{} = \underline{} = \underline{}$$

$$\cot \angle B = \cot 60° = \underline{} = \underline{} = \underline{}$$

$$\sec \angle B = \sec 60° = \underline{} = \underline{} = \underline{}$$

$$\csc \angle B = \csc 60° = \underline{} = \underline{} = \underline{}$$

OBSERVE:

angle A (30°) + angle B (60°) = 90°

and sin 30° = .500

cos 60° = .500

THEREFORE:

the sin of ∠A and the cos of ∠B are EQUAL FUNCTIONS.

CALCULATOR APPLICATION:

Finding the values of the ratio for the basic 6 functions can be simplified with the use of a calculator. Otherwise a set of tables listing the natural trigonometry functions must be available. All the information from these tables is stored in the calculator and makes them very easy to use in determining functions of angles.

EXAMPLE 1: (Function, of given ∠)

Find the tangent of 40°

    a. enter 40

    b. press tan key

    c. READOUT — .83909963

Therefore, the tan 40° = .8391 (Rounded off)

EXAMPLE 2: (Angle, of given function)

Find the angle whose sin is .42262

    a. enter .42262

    b. press invert key

    c. press sin key

    d. READOUT — 25.00011

Therefore, the angle is 25° or the sin of 25° is .42262.

CONVERSION — Degrees to minutes to seconds

    a. Multiply the decimal degree times 60 to get minutes.

    b. Multiply the decimal minutes times 60 to get seconds.

EXAMPLE 3:

Change 33.4869° into min. and sec.

    a. Multiply .4869 times 60 = 29.214 min.

    b. Multiply .214 times 60 = 12.84

Therefore, 33.4869° = 33° 29' 12"

TABLE OF NATURAL TRIGONOMETRIC FUNCTIONS APPLICATION:

An alternative method to using the calculator is the use of the table of natural trigonometric functions.

USE OF TRIGONOMETRY TABLES — NATURAL FUNCTIONS

Values of the functions of angles for each degree and minute from 0°00' to 180°00' are listed in the tables beginning on page 327.

For degrees listed at the top of the page, use the column headings at the top and the minutes column at the left. For degrees listed at the bottom of the page, use the column headings at the bottom, and the minutes column at the right.

EXAMPLE 1:

Find cos 31° 10':

Look up 31° (top of page)

Look up 10' in the minutes column at the left.

Find the value under the "cosine" heading at the top of the page.
cos 31° 10' = .85567

EXAMPLE 2:

Find sec 54° 12':

Look up 54° (bottom of page)

Look up 12' in the minutes column at the right.

Find the value above the "secant" listing at the bottom of the page.
sec 54° 12' = 1.70953

IMPORTANT:

Notice that the "one" is listed only at every 5th minute and is understood to be at those values in between.

INTERPOLATION

When it is necessary to find the function of an angle expressed to seconds, we can determine the proper value through the process of **interpolation**.

EXAMPLE:

To find sin 11° 25′ 40″

First, look up the values of the next highest and next lowest values given in the tables and write them down as below:

sin 11° 26′ 00″	.19823
sin 11° 25′ 40″	
sin 11° 25′ 00″	.19794

Take the difference between the two values:

$$.19823 - .19794 = .00029$$

Set up the following proportion, and solve for x.

$$\frac{40}{60} = \frac{x}{.00029}$$

$$x = .00019$$

Add x to the value for sin 11° 25′ for the final answer.

$$\begin{aligned} \sin 11° 25′ &= .19794 \\ &+ .00019 \\ \sin 11° 25′ 40″ &= .19813 \end{aligned}$$

DECIMAL ANGLES

Fractions of degrees may sometimes be expressed as decimals instead of in minutes and seconds. Functions of decimal angles may be looked up directly in a special table or may be interpolated from a table expressed in minutes.

EXAMPLE:

To find tan 15.45678°

First, select the next highest and next lowest "round" decimals which can be conveniently converted to minutes.

$$\tan 15.50000° = \tan 15°30'00'' = .27732$$
$$\tan 15.45678°$$
$$\tan 15.25000° = \tan 15°15'00'' = .27263$$

Set up the following proportion and solve for x:

$$\frac{(15.45678 - 15.25000)}{(15.50000 - 15.25000)} = \frac{x}{(.27732 - .27263)}$$

$$\frac{.20678}{.25000} = \frac{x}{.00469}$$

$$x = .00388$$

$$\tan 15.25000° = .27263$$
$$+.00388$$
$$\tan 15.45678° = .27651$$

CONCEPT APPLICATIONS:

Set 1

Find the value of functions of the following angles:

1. sin 65° = _____
2. cos 15° = _____
3. tan 32° = _____
4. sec 81° = _____
5. cot 90° = _____
6. sin 36° = _____
7. cos 16° = _____
8. tan 28° = _____
9. cot 10° = _____
10. csc 2° = _____
11. sec 64° = _____
12. tan 13° 5′ = _____
13. sin 16° 31′ = _____
14. cos 18° 5′ = _____
15. tan 38° 39′ = _____
16. csc 1° 2′ = _____
17. tan 89° 32′ = _____
18. csc 5° 55′ = _____
19. sec 21° 39′ = _____
20. sin 67° 49′ = _____
21. cos 55° 41′ = _____
22. cot 72° 21′ = _____
23. tan 1° 15′ = _____
24. csc 13° 59′ = _____
25. tan 42° 25′ = _____

Set 2

Find the angles corresponding to given trigonometric functions:

1. sin .02356 = _____
2. sec 29.899 = _____
3. sec 1.3301 = _____
4. cos .75870 = _____
5. csc 1.7305 = _____
6. tan .7216 = _____
7. sin .11349 = _____
8. cos .94749 = _____
9. tan .50331 = _____
10. cot 1.0006 = _____
11. sec 1.4276 = _____
12. csc 1.0261 = _____
13. cot 4.0611 = _____

14. tan .64117 = _____
15. sec 1.0754 = _____
16. csc 1.1901 = _____
17. cos .88404 = _____
18. sin .05437 = _____
19. csc 1.4142 = _____
20. cot 41.916 = _____
21. cot .01076 = _____
22. tan 1.1674 = _____
23. sec 1.5268 = _____
24. sin .16246 = _____
25. cos .99986 = _____

Set 3

Convert to angles or trigonometric function values as appropriate:

1. sin 1° 15′ 23″ = _____
2. cos 69° 23′ 14″ = _____
3. tan 5° 1′ 2″   = _____
4. csc 89° 18′ 48″ = _____
5. sec 7° 33′ 59″ = _____
6. cot 45° 55′ 18″ = _____
7. sin 55° 41′ 33″ = _____
8. cos 72° 21′ 21″ = _____
9. tan 66° 44′ 44″ = _____
10. csc 77° 38′ 28″ = _____
11. sec 3° 19′ 34″ = _____
12. cot 47° 3′ 54″ = _____
13. sin 15° 37′ 8″ = _____

14. cos 21° 59′ 13″ = _____
15. tan 11° 21′ 4″ = _____
16. csc 4.12194   = _____
17. sec 3.48788   = _____
18. tan .63320   = _____
19. sin .99480   = _____
20. cos .36226   = _____
21. tan .59541   = _____
22. csc 8.70513   = _____
23. sec 13.95848   = _____
24. cot .32733   = _____
25. sin .94622   = _____

UNIT 36

RIGHT TRIANGLE SOLUTIONS

INTRODUCTION

When planning the layout of a job the machinist tries to develop right triangles that will lead to the needed solutions. Solving right triangles involves applying laws of geometry and the functions of trigonometry.

OBJECTIVES:

After completing this unit the student will be able to:

- Solve a right triangle given one side and one acute angle
- Solve a right triangle given two sides

If one side and one acute angle of a right triangle are given, then the remaining sides and angles can be found. Study the following example.

EXAMPLE 1:

Given △ABC: A = 32°15′,   AB = 3.750

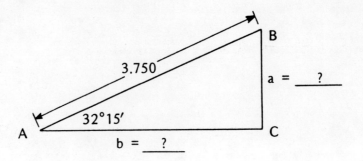

Step 1 Find the measure of the third angle (∠B) (Remember — the sum of the angles = 180° and ∠A + ∠B = 90°).

$$\angle A + \angle B = 90$$
$$\angle B = 90 - \angle A$$
$$\angle B = 90 - 32°15$$
$$\angle B = 57°45$$

Step 2 Find the measure of side BC.

$$\sin \angle A = \frac{BC}{AB} \quad \left(\text{since } \sin \angle A = \frac{opp}{hyp}\right)$$

$$\sin \angle A = \sin 32°15 = .53361 \quad (\text{refer page 343})$$

$$AB = 3.750 \quad (\text{Given in diagram})$$

Substitute these values for sin A and AB in

$$\sin \angle A = \frac{BC}{AB} \quad \text{therefore}$$

$$.53361 = \frac{BC}{3.750}, \quad \text{then cross multiply:}$$

$$(.53361)(3.750) = BC \quad \text{or}$$

$$\text{side } BC = 2.0010 +$$

💡 *NOTE:*

To simplify the form of writing the functions, it is usual practice to let the form "sin ∠A" be expressed as "sin A", omitting the "∠" sign.

Step 3 Find the measure of side AC.

$$\cos A = \frac{AC}{AB} \quad \left(\text{since } \cos A = \frac{adj.}{hyp.}\right)$$

$$\cos A = \cos 32°15' = .84573 \quad \text{(refer page 343)}$$

$$AB = 3.750$$

Substitute and Simplify

$$.84573 = \frac{AC}{3.750}$$

$$\text{side } AC = 3.1714+$$

EXAMPLE 2:

Given △ABC   AC = 9.975   BC = 4.850

Determine: ∠A = ___?___
            ∠B = ___?___
            side AB = ___?___

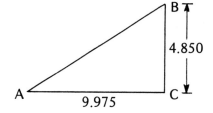

Step 1 $\tan A = \frac{BC}{AC} \quad \left(\text{since } \tan = \frac{opp.}{adj.}\right)$

$$\tan A = .48621$$

$$\angle A = 25°56', \quad \text{therefore} \quad \text{(refer page 339)}$$

$$\angle B = 64°04' \quad \text{complementary } \angle$$

Step 2 Find side AB using the sine of angle A:

$$\sin A = \frac{BC}{AB}$$

$$\sin 25°56' = \frac{4.850}{AB}$$

$$AB = \frac{4.850}{.4373} \quad \text{cross multiply}$$

$$\text{side } AB = 11.0907+$$

CONCEPT APPLICATIONS:

Set 1

Determine the unknown sides:

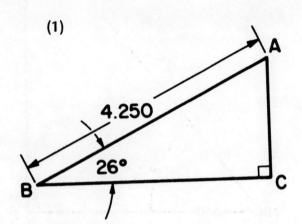

(1) Determine AC = _____
    Determine BC = _____

(2) Determine BC = _____
    Determine AB = _____

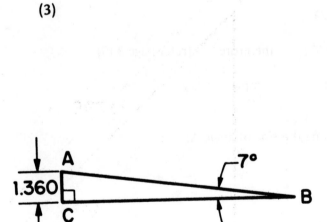

(3) Determine AB = _____
    Determine BC = _____

(4) Determine AB = _____
    Determine AC = _____

(5)

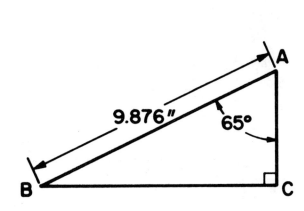

Determine AC = _____ "
Determine BC = _____ "

(6)

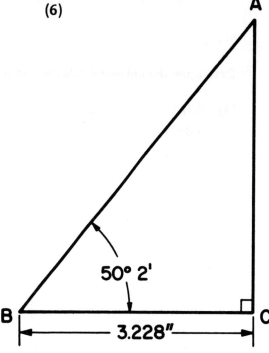

Determine AC = _____ "
Determine AB = _____ "

(7)

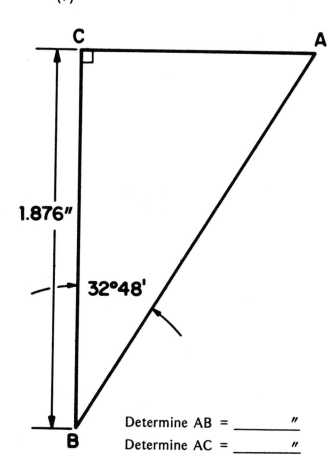

Determine AB = _____ "
Determine AC = _____ "

(8)

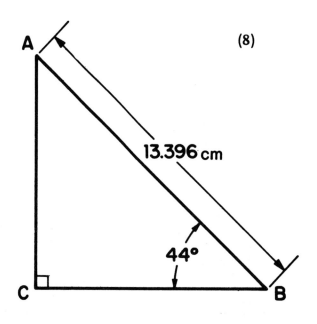

Determine AC = _____ cm
Determine BC = _____ cm

Set 2

Determine the unknown sides and angles.

(1) Angle A = _____
    Side BC = _____ "
    Side AC = _____ "

(2) Angle A = _____
    Side AB = _____ cm
    Side BC = _____ cm

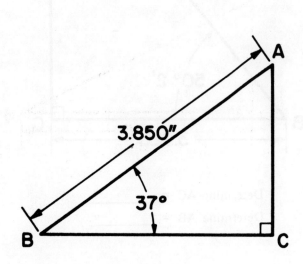
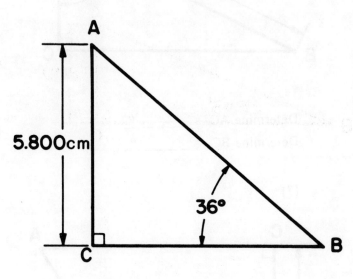

(3) Angle B = _____
    Side AC = _____ "
    Side AB = _____ "

(4) Angle A = _____
    Side AC = _____ mm
    Side AB = _____ mm

(5) Angle BAC = _____
Side AB = _____ "
Side BC = _____ "

(6) Angle CAB = _____
Angle ABC = _____
Side AB = _____ mm

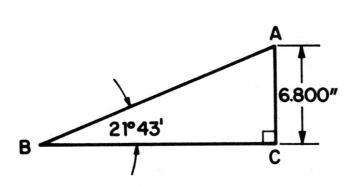
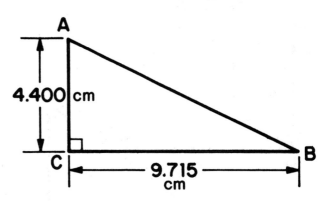

(7) The remaining side = _____ "   (8) The remaining side = _____ mm

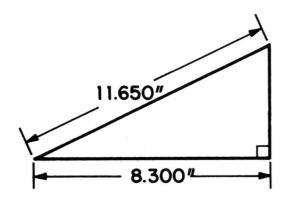
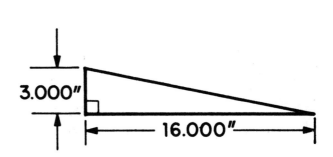

UNIT 37

RIGHT TRIANGLE CONCEPT APPLICATIONS

INTRODUCTION

It is not always obvious how to draw right triangles needed to solve the machining problems. Only through practice and experience can you develop the skill you need to put the known measurements in the form of a right triangle and to use the correct functions to determine the required measurements.

OBJECTIVES:

After completing this unit the student will be able to:

- Use the right triangle to solve distance and angle problems
- Construct auxiliary lines needed to form the right triangles for solving problems

AUXILLARY LINES

Lines which are constructed in addition to given lines for a problem. These lines are usually in the form of a parallel or perpendicular to a given line. They frequently provide the information necessary to solve problems in shop drawings.

Study each of the following drawings to determine information needed in construction of the triangle which solves the problem. Then complete the solutions.

(1) Determine the distance between lines ℓ and m in the following diagram:

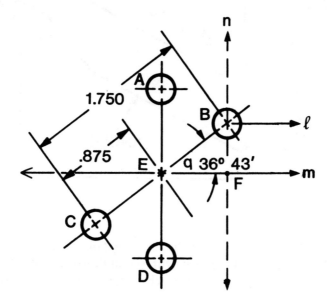

OBSERVE:

Triangle EBF is created by introducing the auxillary line n. Isolate △ EBF as follows:

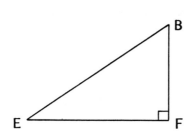

From the given information it is obvious that:

a) side EB = 1.750 − .875

b) ∠q = 36° 43′

c) distance between lines ℓ and m = side BF.

Complete the solution using the sine function for angle q or

$$\sin q = \frac{BF}{BE}$$

BF = _____

279

SINE BAR

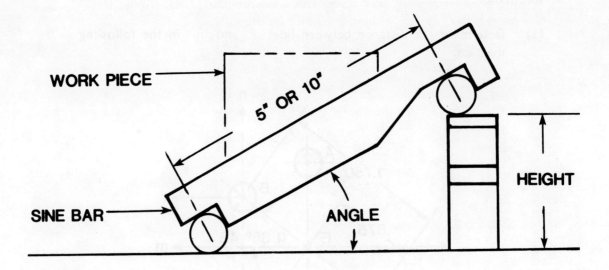

The sine bar is a device used in the shop to hold work at a precise angle. The angle is determined by the height set under one end of the sine bar. The height is the side opposite the angle and the hypotenuse is equal to the length of the sine bar.

Since the sine of an angle is equal to: $\left[\dfrac{opp}{hyp} = sin\right]$

It follows that $\left[hyp \times sin = opp\right]$

Therefore the sine bar setting is equal to the length of the sine bar multiplied by the sine of the desired angle.

(2) What size block should be used under a 10" sine bar to produce an angle of 7°14′27″?

Size Block = _____

(3) What is the angle between a 5" sine bar and a surface plate if size blocks totaling .30679 are placed under one end?

Angle = _____ ° _____ ′ _____ ″

(4) The location of the .605 diameter drill hole and the .255 reamed hole is checked by inserting pins of the same size in the respective holes and then measuring over them with a micrometer. What should the micrometer reading Y be if the holes are properly located?

Distance Y = _____

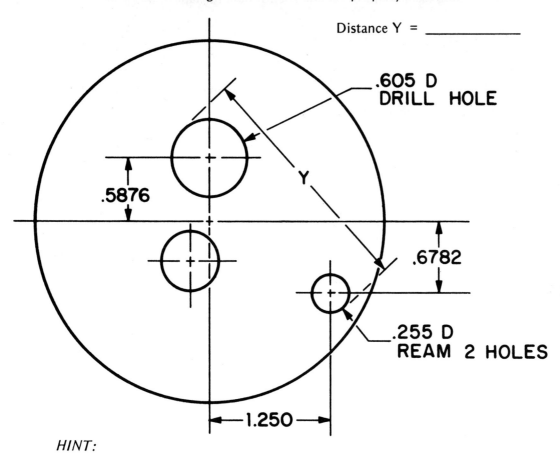

HINT:

First determine the center to center distance of the hypotenuse.

(5) To locate the .355 diameter hole, it is necessary to determine dimensions from the vertical and horizontal centerlines. Determine these dimensions.

Horizontal Distance X = ____

Vertical Distance Y = ____

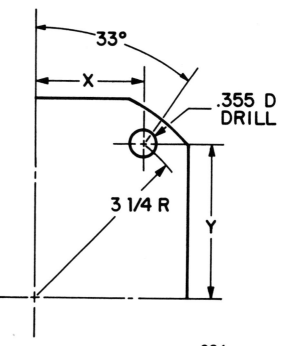

HINT:

The $3\frac{1}{4}$ R is the hypotenuse.

281

TAPERS

Taper is defined as a difference in dimensions at a uniform rate over the length of a part. Taper per foot and taper per inch are the ways of expressing the amount of taper. The formula for tapers:

$$TPF = \frac{D - d}{L} \times 12$$

Where large diameter D is 1.5 inches and small diameter d is .625 inches and the length L is 9 inches:

$$TPF = \frac{D - d}{L} \cdot 12 = \frac{1.5 - .625}{9} \cdot 12 =$$

$$= (.0972) \cdot (12) = 1.166'' \text{ T.P.F.}$$

EXAMPLE:

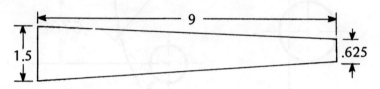

(6)   Distance Y = _____

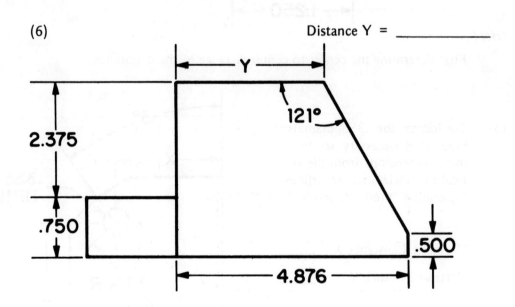

(7) Angle Y = _____

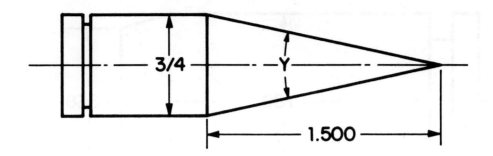

(8) Distance Y = _____

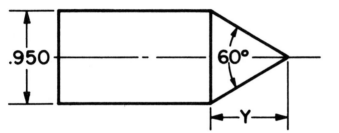

(9) Find the small diameter of the tapered part of the shaft.

Small Diameter Y = _____

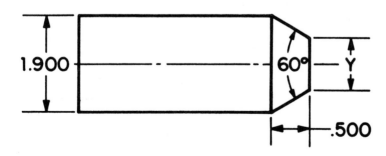

283

(10) Find the included angle X of the tapered part of the shaft.

Angle X = _____

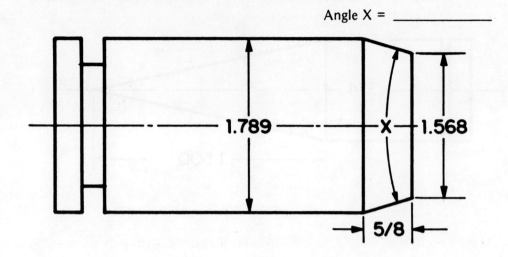

(11) A shaft tapers $\frac{3''}{4}$ per foot. What is the included angle of the taper?

Included Angle = _____

(12) Find the taper per foot of a taper of 8° 27' included angle.

Taper per foot = _____

(13) The width of the flat at the top and bottom of an American Standard Form thread is equal to $\frac{1}{8}$ th of the pitch of that thread. What is the depth of a thread that has a pitch of $\frac{1}{16}$. Note: **The pitch of a thread is equal to one over the number of threads and is the measure from the point on one thread to the corresponding point on the next thread.**

(14) The Acme thread has a depth of .010" plus $\frac{1}{2}$ the pitch, and the included angle between threads is 29°. Find the length of the slope of a $\frac{1}{8}$" pitch thread.

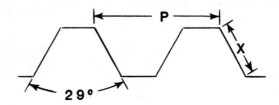

(15) Find the depth of a national Form thread if the pitch were $\frac{1}{13}$". Note: Allow $\frac{1}{8}$P width of flat for root and crest.

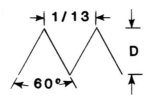

(16) How far does the upper piece rise when the wedge moves ahead .400"?

Distance X = _____

What is the height of a triangle .400 long and 28°?

_____

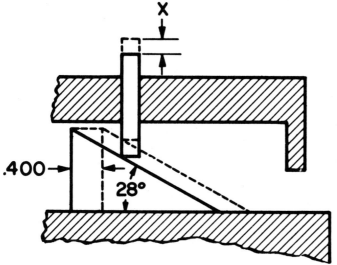

285

(17) In laying out a piece to be machined it is often necessary to determine the angular displacement of various holes from the vertical and horizontal centerlines already established. Determine the angle X.

Angle X = _____

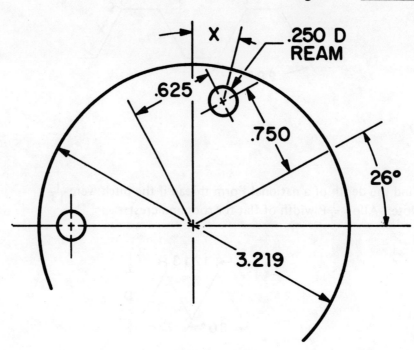

(18) The 6 holes in the template are each $\frac{1''}{8}$ in diameter. Pins of the same size are placed in holes A, B and C. Find the measurements GH and JK which are taken over the pins.

NOTE:

The six holes are equally spaced.

Distance GH = _____

Distance JK = _____

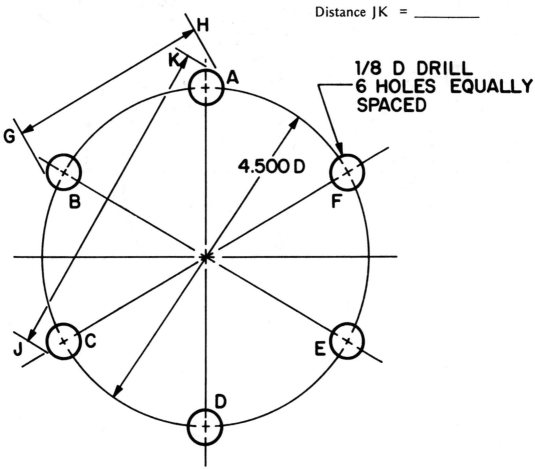

(19) A cylinder barrel is to be laid out with 12 holes equally spaced on a circle 12'' in diameter. Calculate the distance between centers of adjacent holes so that they can be laid out with dividers.

Distance = _____

(20) Find the distance X thru which the cutter would have to move to mill the periphery of the pad ZZ.

Distance X = _____

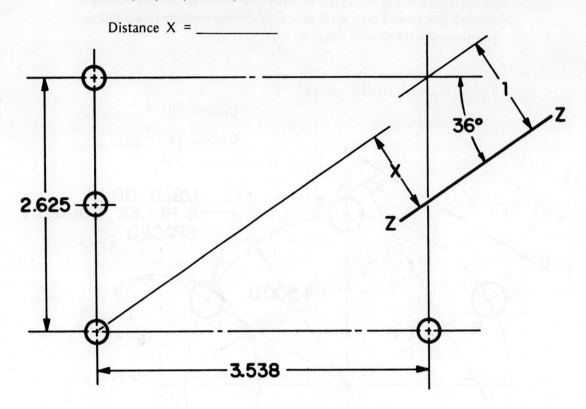

(21) Find the depth X of the $1\frac{5}{16}''$ diameter hole.

Depth X = _____

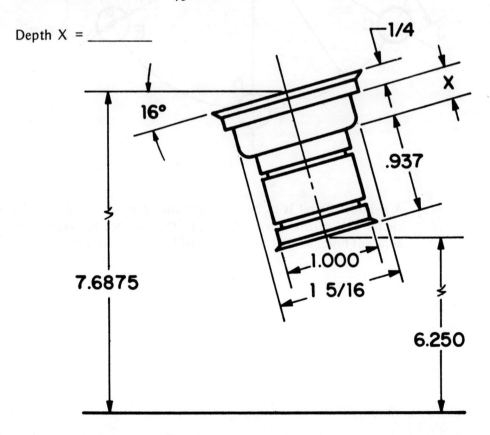

288

(22) Using basic dimensions determine whether there would be a step where the $\frac{1}{4}''$ diameter drill angular hole breaks through the $\frac{1}{4}''$ diameter x $2\frac{1}{16}''$ deep hole.

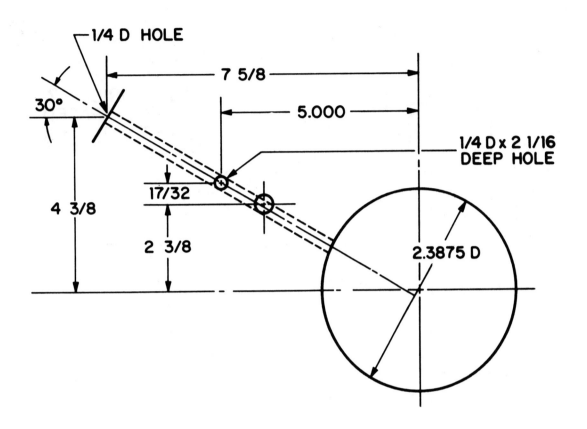

(23) Find Angle X = _____

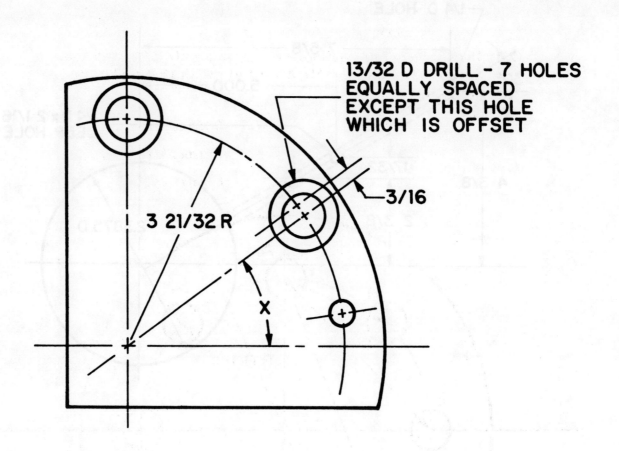

(24) What would be the size of the largest octagon (across flats) that could be milled from a piece of smooth stock 2" in diameter?

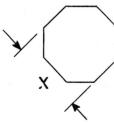

Size = _____

(25) The diameter of the circle locating these 8 equally spaced drill holes is 8". Find the distances AH, AG and AF, which are needed to check the location of the holes.

Distance AH = _____

Distance AG = _____

Distance AF = _____

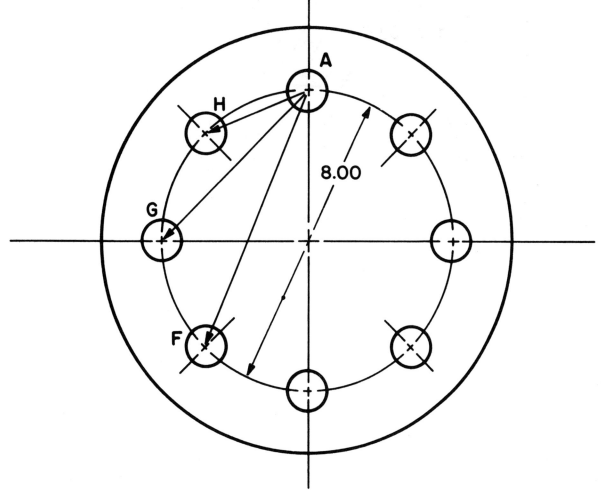

The radius becomes one of the equal sides of an isosoles triangle.

291

(26) To check the location of the $\frac{5''}{8}$ diameter drill angular hole the distance X must be calculated. What is that distance?

Note that a $\frac{5''}{8}$ diameter pin is used in the hole.

Distance X = _____

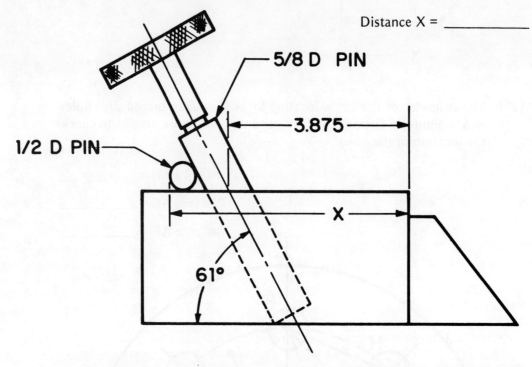

(27) After machining a groove in a piston, the depth is checked by measuring the distance over pins placed in the groove as shown. If the pins are .136" in diameter and the diameter of the piston is 5.375", find the measurement X over pins.

Measurement X = _____

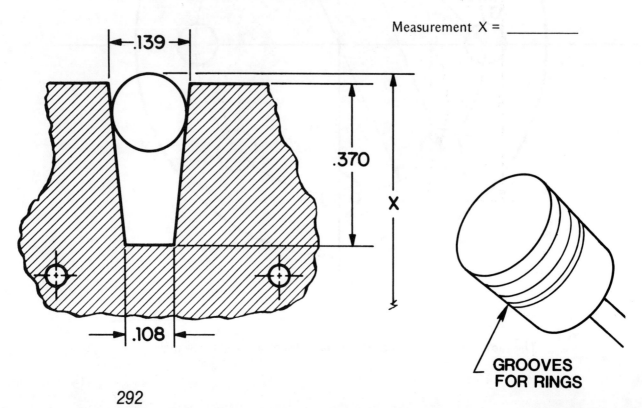

(28) What will the measurement X be if the five sides of this symmetrical figure are each 2 inches in length?

Measurement X = _____

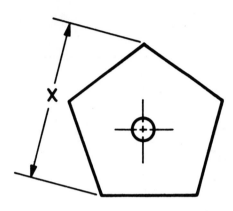

(29) Find the distance X on this symmetrical template.

Distance X = _____

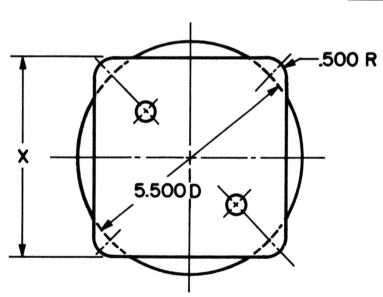

(30) If two disks $1\frac{1}{2}''$ and $1\frac{7}{8}''$ in diameter respectively are in contact with each other, what is the included angle between the straight edges?

Included Angle = _____

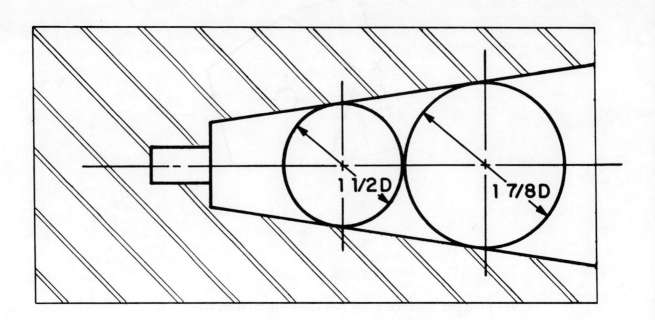

(31) Two disks, one $\frac{7}{8}''$ and the other $2\frac{1}{4}''$ in diameter, are used to measure a taper. The distance between centers of the disks is $3\frac{7}{8}''$. What is the rate of taper per foot?

Taper per foot = _____

(32) In order to machine the $\frac{5}{8}''$ R as shown, it is necessary to compute the angle from the vertical centerline to the center of the $\frac{5}{8}''$ R. What is the size of that angle?

Angle = _____

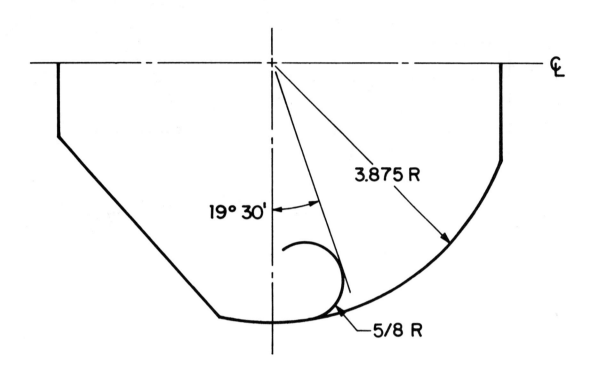

(33) If the disks in problem #30 had been 4" apart on centers, what would the angle between the straight edges have been?

Included Angle = _____

(34) How far apart on centers should two standard reference disks 1.250" and 2" in diameter respectively be placed to layout an angle of 9° 35'?

Center Distance = _____

(35) What is the size of the included angle X on the torque rod shown below?

Angle X = _____

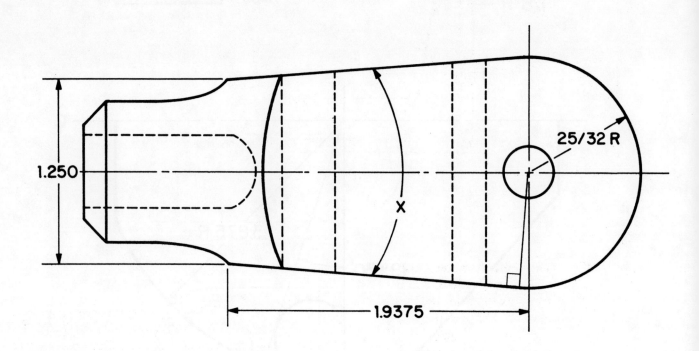

(36) Angle B = _____

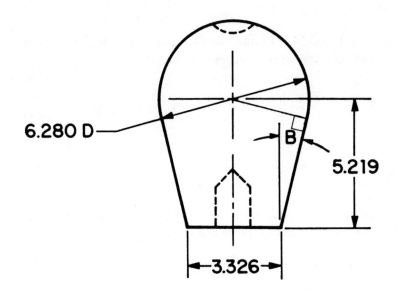

296

(37) Distance X = _____

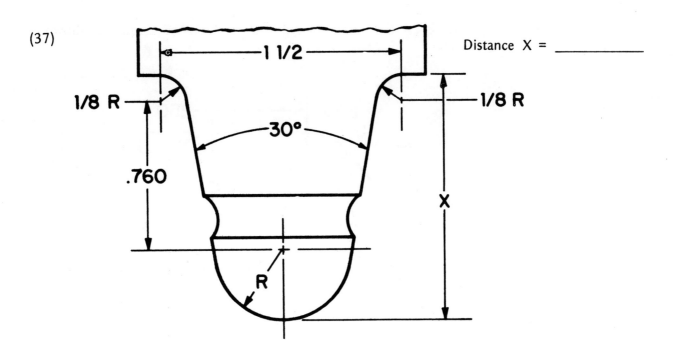

(38) The 1 3/8" diameter must be milled and made to end perfectly with the .875 diameter. To do this, it is necessary to calculate the angle X thru which the milling cutter must revolve. What is the size of angle X?

Angle X = _____

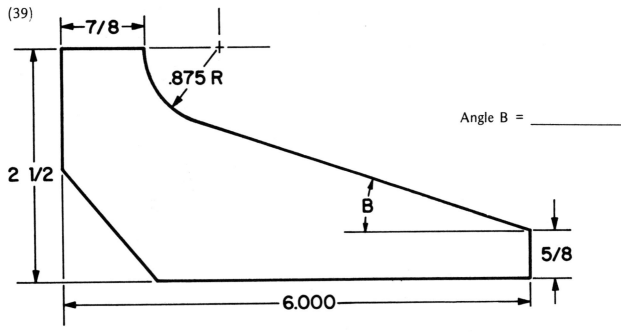

Angle B = _____

297

(40) Angle B = _____

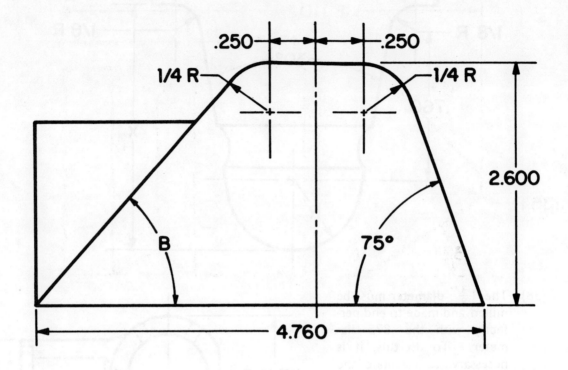

(41) Distance X = _____

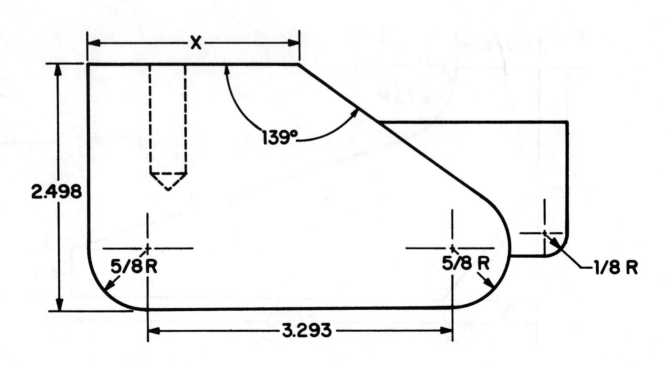

298

(42)   Distance X = _____

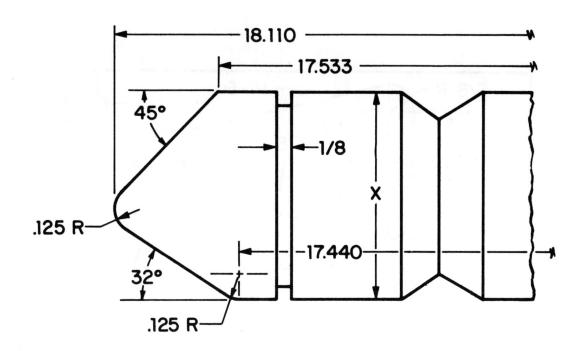

(43)   Distance X = _____

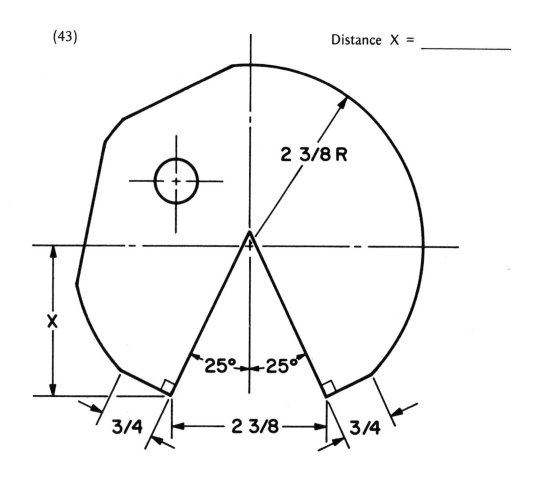

299

(44) Angle B = _____

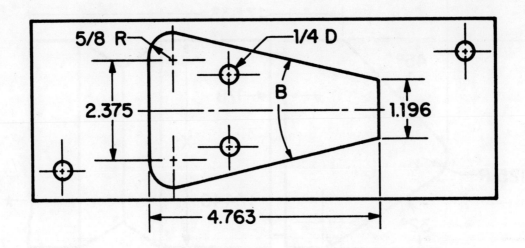

(45) Distance X = _____

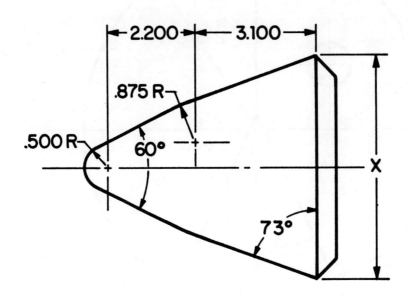

300

(46) Find distance X on the punch which allows .010 cutting clearance between punch and die.

Distance X = _____

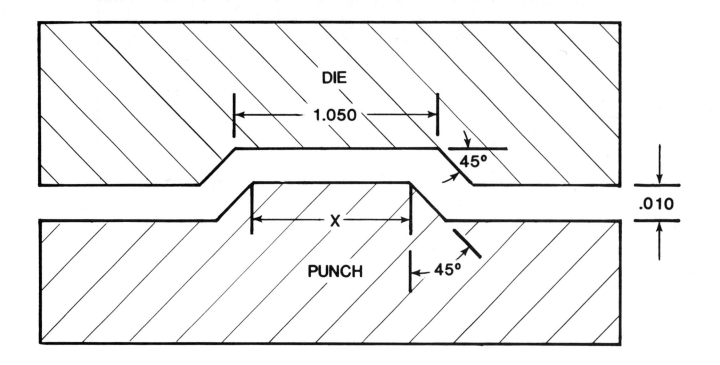

(47) Angle I = _____

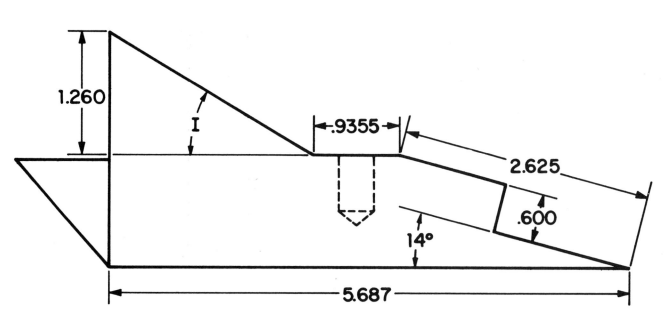

301

(48) Dimension C = _____

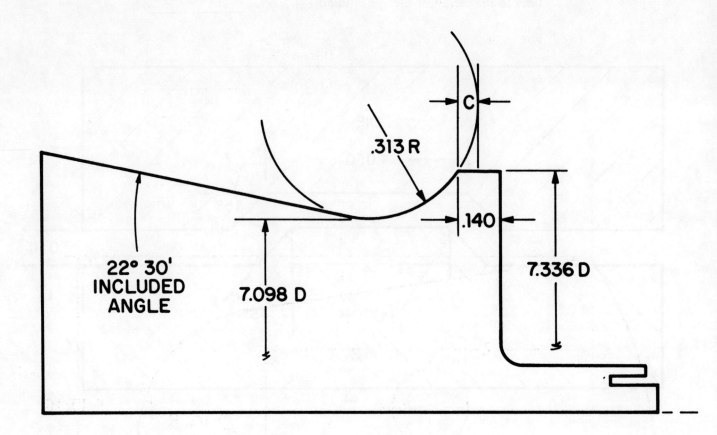

(49) To machine this counterweight, it is necessary to compute:

(a) The linear dimensions from the vertical and horizontal centerlines to the center of the $1\frac{1}{2}"$ Radius.

(b) The angle X.

Distance Z = _____

Distance Y = _____

Angle X = _____

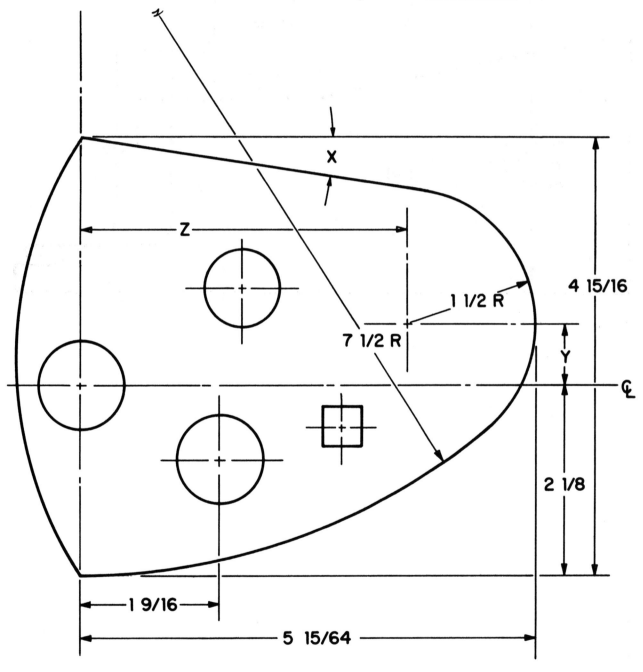

(50) Two parallel lines .0608 apart are to be connected by two radii as shown by the sketch. If the locating distance to the meeting point of the two curves is 1.250", find the locating distances X and Y to the center of the two respective radii.

Distance X = _____

Distance Y = _____

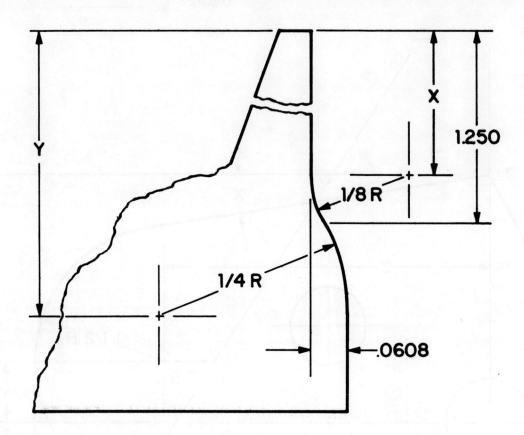

(51) Find the angle X at which the slot shown in the section view is machined.

Angle X = _____

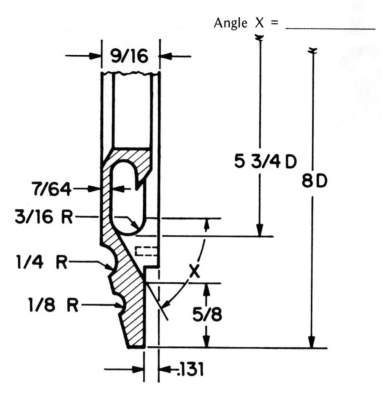

(52) Find the distance X which is necessary to check the location of the 1" formed radius and the 10° angle.

Distance X = _____

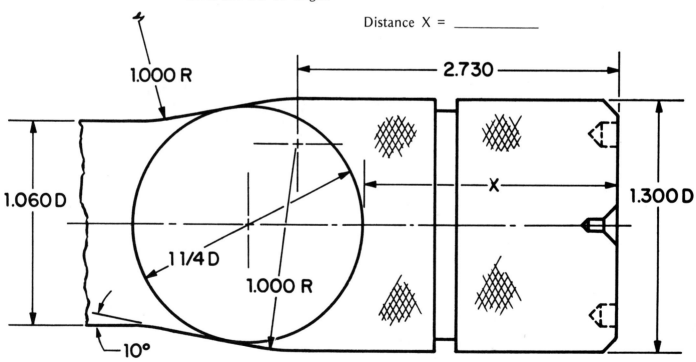

UNIT 38

LAW OF SINES

INTRODUCTION

One law used in trigonometry is stated in terms of the sines and sides of angles when a triangle is not a right triangle. This is called an oblique triangle. The law of sines must be applied to solve oblique triangle problems.

OBJECTIVE:

After completing this unit the student will be able to:

- Solve for sides and angles of oblique triangles using the Law of Sines

✋ OBLIQUE TRIANGLE

Any triangle which does not contain a right angle. To determine the measures for the angles and sides of an OBLIQUE triangle, we need additional rules. One of the most useful is the Law of Sines.

💡 LAW OF SINES:

In either triangle ABC, the altitude, "h" is introduced as an auxiliary segment. The result is obvious in the form of two right triangles. ($\triangle$ACD and $\triangle$BCD).

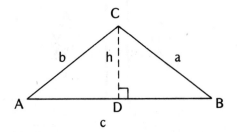 or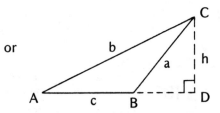

In both triangles, $\longrightarrow \dfrac{h}{b} = \sin A$

In the first triangle, $\longrightarrow \dfrac{h}{a} = \sin B$

If we solve both equations for "h":

$\dfrac{h}{b} = \sin A$ $\qquad\qquad\qquad$ $\dfrac{h}{a} = \sin B$

$\therefore h = b \sin A$ $\qquad\qquad\qquad$ $\therefore h = a \sin B$

since $h = h$ and substitute

we have $(a \sin B = b \sin A)$

or $\qquad \dfrac{a}{b} = \dfrac{\sin A}{\sin B}$

In the same way we can draw perpendiculars from the vertices A and B to the opposite sides to obtain the following relations:

and
$$\dfrac{b}{c} = \dfrac{\sin B}{\sin C}$$
$$\dfrac{a}{c} = \dfrac{\sin A}{\sin C}$$

💡 This relation between the sides and the sines of the opposite angles is called the LAW OF SINES and may be expressed as follows:

"The sides of a triangle are proportional to the sines of the opposite angles."

If we multiply $\dfrac{a}{b} = \dfrac{\sin A}{\sin B}$ by b, and divide by $\sin A$, we have

$$\dfrac{a}{\sin A} = \dfrac{b}{\sin B}$$

Similarly, we may obtain the following:

1. $\dfrac{a}{\sin A} = \dfrac{b}{\sin B} = \dfrac{c}{\sin C}$ or

2. $a \sin B = b \sin A$
   $a \sin C = c \sin A$
   $b \sin C = c \sin B$

FORMS OF THE LAW OF SINES

To use the LAW OF SINES to solve a proportion, three of the four parts must be known:

                      2 sines and 1 side or
                      1 sine and 2 sides

EXAMPLE:

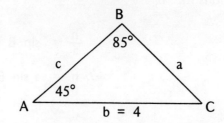

Given: ∠A = 45°, ∠B = 85°, b = 4

Find: ∠C, side a, and side c.

Solution for C =   ?

$$C = 180° - (85° + 45°)$$
$$C = 180° - 130°$$
$$C = 50°$$

Solution for side a =   ?

$\dfrac{b}{\sin B} = \dfrac{a}{\sin A}$	Law of Sines
$\dfrac{4}{\sin 85°} = \dfrac{a}{\sin 45°}$	Substitution
$\dfrac{4}{.99619} = \dfrac{a}{.70711}$	Substitution from tables
$.99619\, a = 2.82844$	Cross Multiplication
$a = \dfrac{2.82844}{.99619}$	Division Prop. of =
$\therefore a = 2.839$	

Solution for side c = __?__

$$\frac{b}{\sin B} = \frac{c}{\sin C}$$ Laws of Sines

$$\frac{4}{\sin 85°} = \frac{c}{\sin 50°}$$ Substitution

$$\frac{4}{.99619} = \frac{c}{.76604}$$ Substitution from tables

$$.99619\, c = 3.06416$$ Cross Multiplication

$$c = \frac{3.06416}{.99619}$$ Division Prop. of =

$$\therefore c = 3.0758$$

CONCEPT APPLICATIONS:

Problem 1

Solve for X = _____

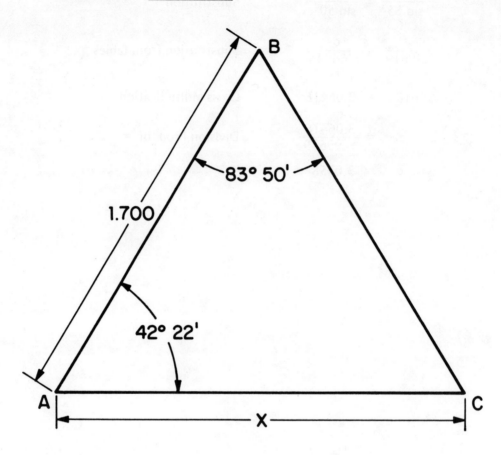

Problem 2

Solve for X = _____ cm

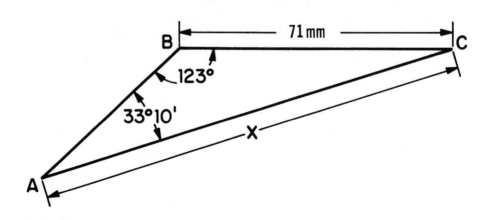

310

Problem 3

Solve for side X = _____

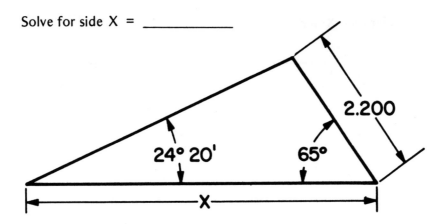

Problem 4

Solve for side X = _____ mm

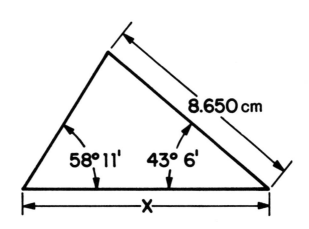

Problem 5

Determine X and Z.

X = _____

Z = _____

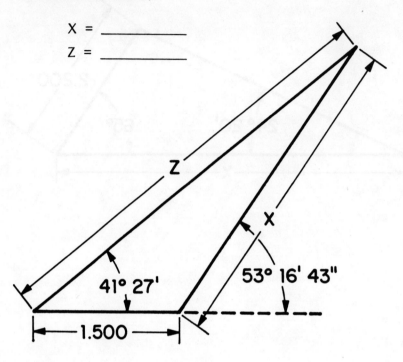

Problem 6

Solve for side X = _____

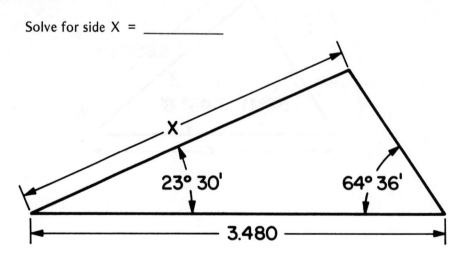

Problem 7

Solve for Angle B and Distance X.

∠ B = _____

X = _____ mm

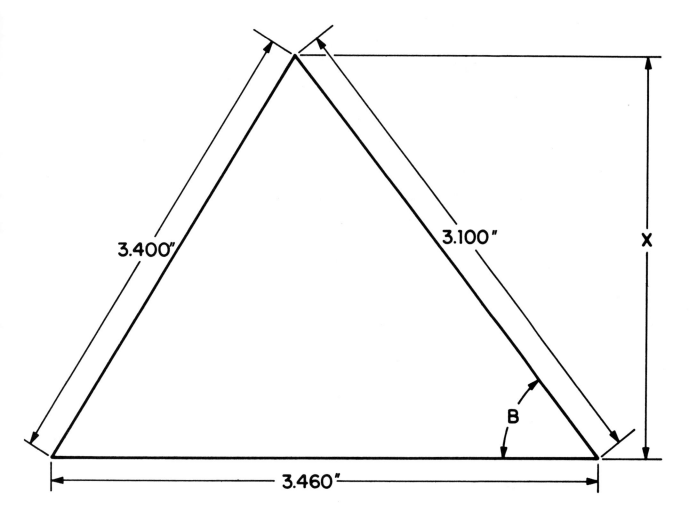

Problem 8

Solve for side X and angle B.

X = _____

∠B = _____

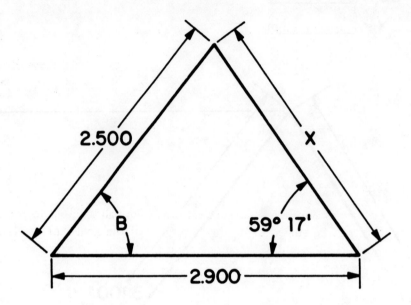

Problem 9

Determine X and Z.

X = _____ m

Z = _____ m

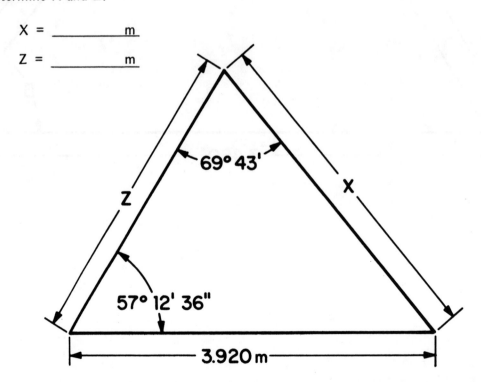

UNIT 39

LAW OF COSINES

INTRODUCTION

The law of cosines is another special equation which can be used to solve triangles that cannot be solved using the functions of the angles or the law of sines. When two sides and the included angle are given, a form of the law of cosines must be applied.

OBJECTIVE:

After completing this unit the student will be able to:

- Solve for sides and angles of oblique triangles using the Law of Cosines

LAW OF COSINES:

This law gives the value of one side of a triangle in terms of the other two sides and the angle included between them.

GIVEN: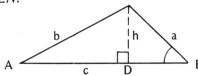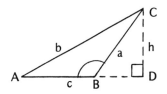

In **both** figures,	$a^2 = h^2 + (BD)^2$
In the **first** figure,	$BD = c - AD$
In the **second** figure,	$BD = AD - c$
In **both** figures,	$(BD)^2 = (AD)^2 - 2c \cdot AD + c^2$

315

Therefore, in all cases,

$$a^2 = h^2 + (AD)^2 + c^2 - 2c \cdot AD$$

but, $h^2 + (AD)^2 = b^2$

and, $AD = b \cos A$

Therefore,: $[a^2 = b^2 + c^2 - 2bc \cdot \cos A]$

Similar proofs give:

$[b^2 = c^2 + a^2 - 2ca \cdot \cos B]$

$[c^2 = a^2 + b^2 - 2ab \cdot \cos C]$

FORMS OF THE LAW OF COSINES ALTERNATES TO SOLVE FOR SIDES

These three formulas have precisely the same form and the LAW OF COSINES may be stated as follows:

> The square of any side of a triangle is equal to the sum of the squares of the other two sides minus twice their product times the cosine of the included angle.

By transposing terms and cross multiplying factors we may write these three rules in terms of the cosines of the three angles:

1. $\cos A = \dfrac{b^2 + c^2 - a^2}{2bc}$

2. $\cos B = \dfrac{a^2 + c^2 - b^2}{2ac}$

3. $\cos C = \dfrac{a^2 + b^2 - c^2}{2ab}$

LAW OF COSINES ALTERNATES TO SOLVE FOR ANGLES

EXAMPLE:

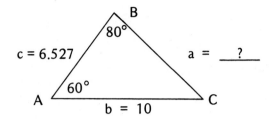

GIVEN: Triangle ABC, with ∠A = 60, ∠B = 80, b = 10, c = 6.527

Find: ∠C = _____ and

a = _____

SOLUTION for ∠C:

∠C = 180° − (∠A + ∠B)
∠C = 180° − (60° + 80°)
∠C = 180° − 140°
∠C = 40°

SOLUTION for a:

$a^2 = b^2 + c^2 - 2bc \cdot \cos A$
$a^2 = (10)^2 + (6.527)^2 - 2(10)(6.527) \cos 60°$
$a^2 = 100 + 42.60 - 2(65.27)(.5)$
$a^2 = 142.60 - 65.27$
$a^2 = 77.33$
$a = \sqrt{77.33}$
a = 8.7937 or 8.794

CONCEPT APPLICATIONS:

Problem 1

Solve for Angle O = _____

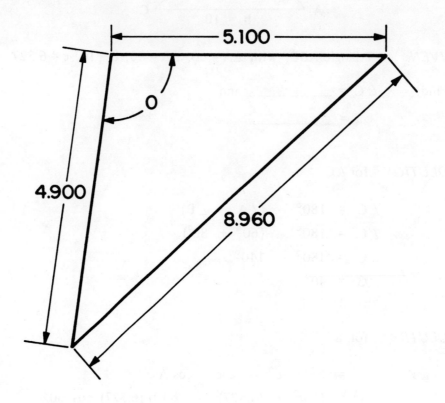

Problem 2

Solve for Angle B = _____

Solve for Side X = _____

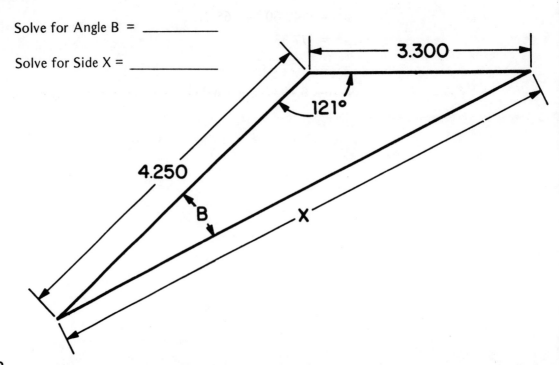

Problem 3

Solve for Angle A = _____

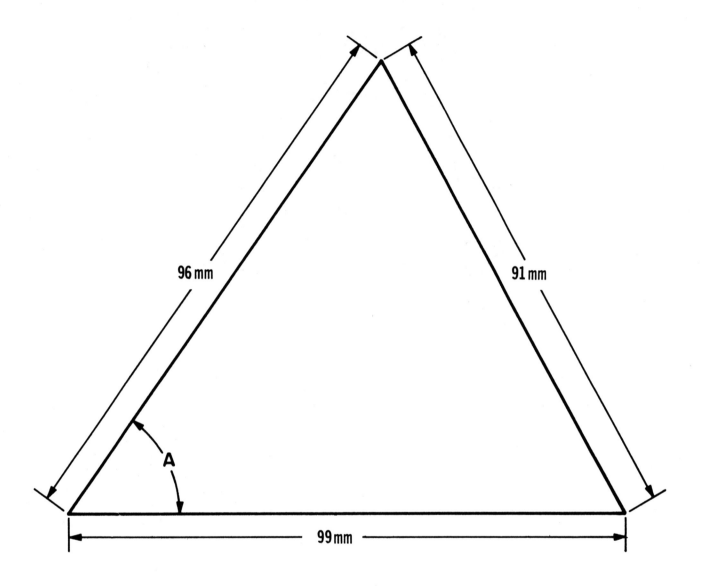

Problem 4

Solve for Diameter X = _____

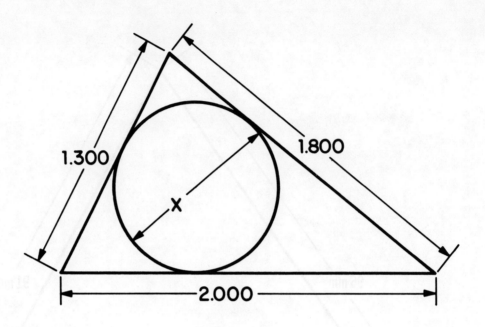

Problem 5

Solve for Angle A = _____

Solve for Angle B = _____

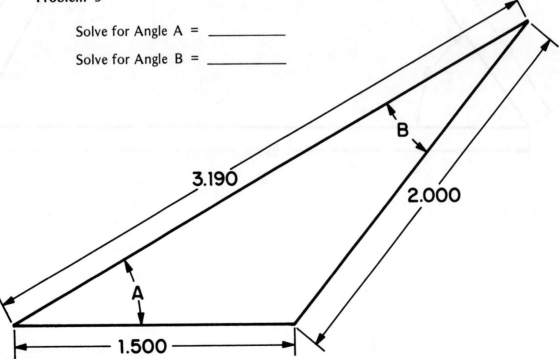

Problem 6

Solve for Angle A = _____

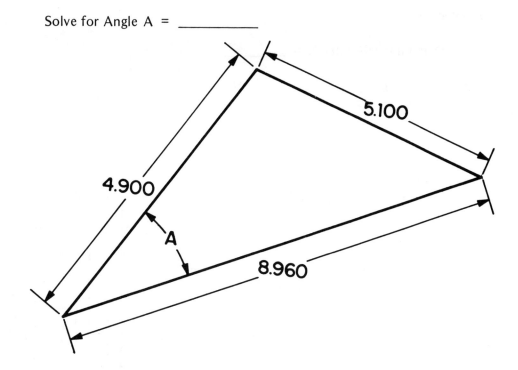

Problem 7

Solve for Angle B = _____

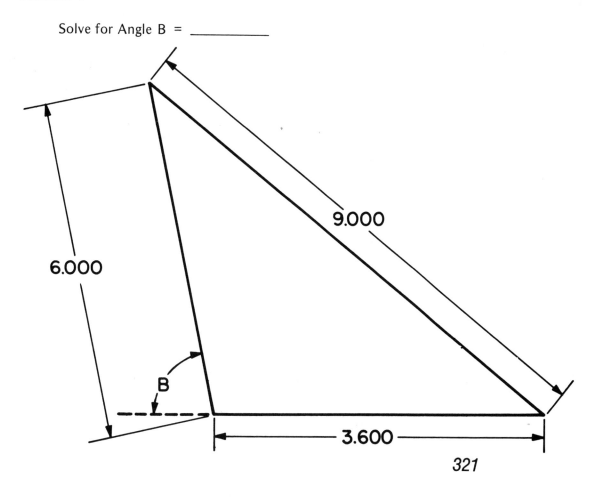

Problem 8

Determine Diameter X = _____

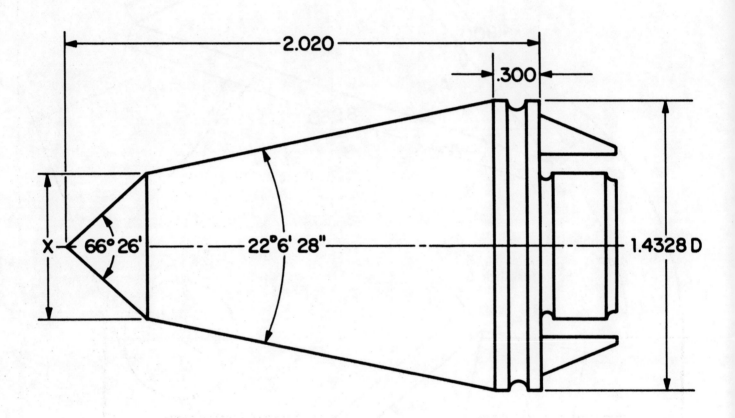

322

Problem 9

Determine Distance X = _____

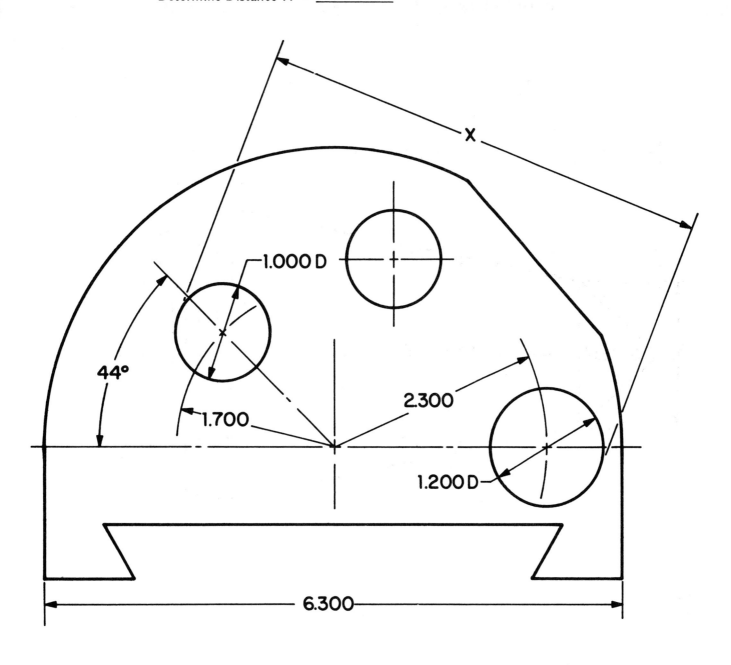

Problem 10

Determine Diameter X = _____

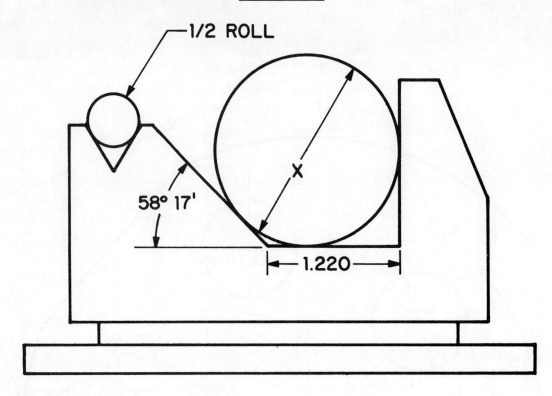

Problem 11

Determine Diameter X and Length Y = _____

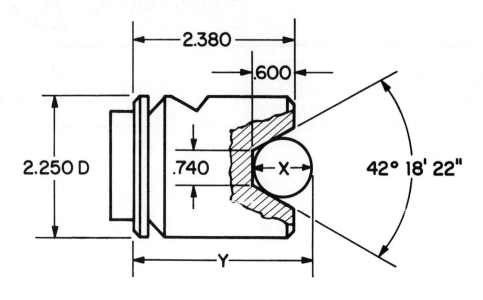

TABLES OF NATURAL TRIGONOMETRIC FUNCTIONS AND SQUARES AND SQUARE ROOTS

TABLES OF NATURAL TRIGONOMETRIC FUNCTIONS

0°→	sin	csc	tan	cot	sec	cos	←179°
0	0.00000	∞	0.00000	∞	1.00000	1.00000	60
1	.00029	3437.75	.00029	3437.75	.00000	.00000	59
2	.00058	1718.87	.00058	1718.87	.00000	.00000	58
3	.00087	1145.92	.00087	1145.92	.00000	.00000	57
4	.00116	859.437	.00116	859.436	.00000	.00000	56
5	0.00145	687.550	0.00145	687.549	1.00000	1.00000	55
6	.00175	572.958	.00175	572.957	.00000	.00000	54
7	.00204	491.107	.00204	491.106	.00000	.00000	53
8	.00233	429.719	.00233	429.718	.00000	.00000	52
9	.00262	381.972	.00262	381.971	.00000	.00000	51
10	0.00291	343.775	0.00291	343.774	1.00000	1.00000	50
11	.00320	312.523	.00320	312.521	.00001	0.99999	49
12	.00349	286.479	.00349	286.478	.00001	.99999	48
13	.00378	264.443	.00378	264.441	.00001	.99999	47
14	.00407	245.554	.00407	245.552	.00001	.99999	46
15	0.00436	229.184	0.00436	229.182	1.00001	0.99999	45
16	.00465	214.860	.00465	214.858	.00001	.99999	44
17	.00495	202.221	.00495	202.219	.00001	.99999	43
18	.00524	190.987	.00524	190.984	.00001	.99999	42
19	.00553	180.935	.00553	180.932	.00002	.99998	41
20	0.00582	171.888	0.00582	171.885	1.00002	0.99998	40
21	.00611	163.703	.00611	163.700	.00002	.99998	39
22	.00640	156.262	.00640	156.259	.00002	.99998	38
23	.00669	149.468	.00669	149.465	.00002	.99998	37
24	.00698	143.241	.00698	143.237	.00002	.99998	36
25	0.00727	137.511	0.00727	137.507	1.00003	0.99997	35
26	.00756	132.222	.00756	132.219	.00003	.99997	34
27	.00785	127.325	.00785	127.321	.00003	.99997	33
28	.00814	122.778	.00815	122.774	.00003	.99997	32
29	.00844	118.544	.00844	118.540	.00004	.99996	31
30	0.00873	114.593	0.00873	114.589	1.00004	0.99996	30
31	.00902	110.897	.00902	110.892	.00004	.99996	29
32	.00931	107.431	.00931	107.426	.00004	.99996	28
33	.00960	104.176	.00960	104.171	.00005	.99995	27
34	.00989	101.112	.00989	101.107	.00005	.99995	26
35	0.01018	98.2230	0.01018	98.2179	1.00005	0.99995	25
36	.01047	95.4947	.01047	95.4895	.00005	.99995	24
37	.01076	92.9139	.01076	92.9085	.00006	.99994	23
38	.01105	90.4689	.01105	90.4633	.00006	.99994	22
39	.01134	88.1492	.01135	88.1436	.00006	.99994	21
40	0.01164	85.9456	0.01164	85.9398	1.00007	0.99993	20
41	.01193	83.8495	.01193	83.8435	.00007	.99993	19
42	.01222	81.8532	.01222	81.8470	.00007	.99993	18
43	.01251	79.9497	.01251	79.9434	.00008	.99992	17
44	.01280	78.1327	.01280	78.1263	.00008	.99992	16
45	0.01309	76.3966	0.01309	76.3900	1.00009	0.99991	15
46	.01338	74.7359	.01338	74.7292	.00009	.99991	14
47	.01367	73.1458	.01367	73.1390	.00009	.99991	13
48	.01396	71.6221	.01396	71.6151	.00010	.99990	12
49	.01425	70.1605	.01425	70.1533	.00010	.99990	11
50	0.01454	68.7574	0.01455	68.7501	1.00011	0.99989	10
51	.01483	67.4093	.01484	67.4019	.00011	.99989	9
52	.01513	66.1130	.01513	66.1055	.00011	.99989	8
53	.01542	64.8657	.01542	64.8580	.00012	.99988	7
54	.01571	63.6646	.01571	63.6567	.00012	.99988	6
55	0.01600	62.5072	0.01600	62.4992	1.00013	0.99987	5
56	.01629	61.3911	.01629	61.3829	.00013	.99987	4
57	.01658	60.3141	.01658	60.3058	.00014	.99986	3
58	.01687	59.2743	.01687	59.2659	.00014	.99986	2
59	.01716	58.2698	.01716	58.2612	.00015	.99985	1
60	0.01745	57.2987	0.01746	57.2900	1.00015	0.99985	0
90°→	cos	sec	cot	tan	csc	sin	←89°

1°→	sin	csc	tan	cot	sec	cos	←178°
0	0.01745	57.2987	0.01746	57.2900	1.00015	0.99985	60
1	.01774	56.3595	.01775	56.3506	.00016	.99984	59
2	.01803	55.4505	.01804	55.4415	.00016	.99984	58
3	.01832	54.5705	.01833	54.5613	.00017	.99983	57
4	.01862	53.7179	.01862	53.7086	.00017	.99983	56
5	0.01891	52.8916	0.01891	52.8821	1.00018	0.99982	55
6	.01920	52.0903	.01920	52.0807	.00018	.99982	54
7	.01949	51.3129	.01949	51.3032	.00019	.99981	53
8	.01978	50.5584	.01978	50.5485	.00020	.99980	52
9	.02007	49.8258	.02007	49.8157	.00020	.99980	51
10	0.02036	49.1141	0.02036	49.1039	1.00021	0.99979	50
11	.02065	48.4224	.02066	48.4121	.00021	.99979	49
12	.02094	47.7500	.02095	47.7395	.00022	.99978	48
13	.02123	47.0960	.02124	47.0853	.00023	.99977	47
14	.02152	46.4596	.02153	46.4489	.00023	.99977	46
15	0.02181	45.8403	0.02182	45.8294	1.00024	0.99976	45
16	.02211	45.2372	.02211	45.2261	.00024	.99976	44
17	.02240	44.6498	.02240	44.6386	.00025	.99975	43
18	.02269	44.0775	.02269	44.0661	.00026	.99974	42
19	.02298	43.5196	.02298	43.5081	.00026	.99974	41
20	0.02327	42.9757	0.02328	42.9641	1.00027	0.99973	40
21	.02356	42.4452	.02357	42.4335	.00028	.99972	39
22	.02385	41.9277	.02386	41.9158	.00028	.99972	38
23	.02414	41.4227	.02415	41.4106	.00029	.99971	37
24	.02443	40.9296	.02444	40.9174	.00030	.99970	36
25	0.02472	40.4482	0.02473	40.4358	1.00031	0.99969	35
26	.02501	39.9780	.02502	39.9655	.00031	.99969	34
27	.02530	39.5185	.02531	39.5059	.00032	.99968	33
28	.02560	39.0696	.02560	39.0568	.00033	.99967	32
29	.02589	38.6307	.02589	38.6177	.00034	.99966	31
30	0.02618	38.2016	0.02619	38.1885	1.00034	0.99966	30
31	.02647	37.7818	.02648	37.7686	.00035	.99965	29
32	.02676	37.3713	.02677	37.3579	.00036	.99964	28
33	.02705	36.9695	.02706	36.9560	.00037	.99963	27
34	.02734	36.5763	.02735	36.5627	.00037	.99963	26
35	0.02763	36.1914	0.02764	36.1776	1.00038	0.99962	25
36	.02792	35.8145	.02793	35.8006	.00039	.99961	24
37	.02821	35.4454	.02822	35.4313	.00040	.99960	23
38	.02850	35.0838	.02851	35.0695	.00041	.99959	22
39	.02879	34.7295	.02881	34.7151	.00041	.99959	21
40	0.02908	34.3823	0.02910	34.3678	1.00042	0.99958	20
41	.02938	34.0420	.02939	34.0273	.00043	.99957	19
42	.02967	33.7083	.02968	33.6935	.00044	.99956	18
43	.02996	33.3812	.02997	33.3662	.00045	.99955	17
44	.03025	33.0603	.03026	33.0452	.00046	.99954	16
45	0.03054	32.7455	0.03055	32.7303	1.00047	0.99953	15
46	.03083	32.4367	.03084	32.4213	.00048	.99952	14
47	.03112	32.1337	.03114	32.1181	.00048	.99952	13
48	.03141	31.8362	.03143	31.8205	.00049	.99951	12
49	.03170	31.5442	.03172	31.5284	.00050	.99950	11
50	0.03199	31.2576	0.03201	31.2416	1.00051	0.99949	10
51	.03228	30.9761	.03230	30.9599	.00052	.99948	9
52	.03257	30.6996	.03259	30.6833	.00053	.99947	8
53	.03286	30.4280	.03288	30.4116	.00054	.99946	7
54	.03316	30.1612	.03317	30.1446	.00055	.99945	6
55	0.03345	29.8990	0.03346	29.8823	1.00056	0.99944	5
56	.03374	29.6414	.03376	29.6245	.00057	.99943	4
57	.03403	29.3881	.03405	29.3711	.00058	.99942	3
58	.03432	29.1392	.03434	29.1220	.00059	.99941	2
59	.03461	28.8944	.03463	28.8771	.00060	.99940	1
60	0.03490	28.6537	0.03492	28.6363	1.00061	0.99939	0
91°→	cos	sec	cot	tan	csc	sin	←88°

2°→	sin	csc	tan	cot	sec	cos ←177°	
′							′
0	0.03490	28.6537	0.03492	28.6363	1.00061	0.99939	60
1	.03519	.4170	.03521	.3994	.00062	.99938	59
2	.03548	28.1842	.03550	28.1664	.00063	.99937	58
3	.03577	27.9551	.03579	27.9372	.00064	.99936	57
4	.03606	.7298	.03609	.7117	.00065	.99935	56
5	0.03635	27.5080	0.03638	27.4899	1.00066	0.99934	55
6	.03664	.2898	.03667	.2715	.00067	.99933	54
7	.03693	27.0750	.03696	27.0566	.00068	.99932	53
8	.03723	26.8636	.03725	26.8450	.00069	.99931	52
9	.03752	.6555	.03754	.6367	.00070	.99930	51
10	0.03781	26.4505	0.03783	26.4316	1.00072	0.99929	50
11	.03810	.2487	.03812	.2296	.00073	.99927	49
12	.03839	26.0499	.03842	26.0307	.00074	.99926	48
13	.03868	25.8542	.03871	25.8348	.00075	.99925	47
14	.03897	.6613	.03900	.6418	.00076	.99924	46
15	0.03926	25.4713	0.03929	25.4517	1.00077	0.99923	45
16	.03955	.2841	.03958	.2644	.00078	.99922	44
17	.03984	25.0997	.03987	25.0798	.00079	.99921	43
18	.04013	24.9179	.04016	24.8978	.00081	.99919	42
19	.04042	.7387	.04046	.7185	.00082	.99918	41
20	0.04071	24.5621	0.04075	24.5418	1.00083	0.99917	40
21	.04100	.3880	.04104	.3675	.00084	.99916	39
22	.04129	.2164	.04133	.1957	.00085	.99915	38
23	.04159	24.0471	.04162	24.0263	.00087	.99913	37
24	.04188	23.8802	.04191	23.8593	.00088	.99912	36
25	0.04217	23.7156	0.04220	23.6945	1.00089	0.99911	35
26	.04246	.5533	.04250	.5321	.00090	.99910	34
27	.04275	.3932	.04279	.3718	.00091	.99909	33
28	.04304	.2352	.04308	.2137	.00093	.99907	32
29	.04333	23.0794	.04337	23.0577	.00094	.99906	31
30	0.04362	22.9256	0.04366	22.9038	1.00095	0.99905	30
31	.04391	.7739	.04395	.7519	.00097	.99904	29
32	.04420	.6241	.04424	.6020	.00098	.99902	28
33	.04449	.4764	.04454	.4541	.00099	.99901	27
34	.04478	.3305	.04483	.3081	.00100	.99900	26
35	0.04507	22.1865	0.04512	22.1640	1.00102	0.99898	25
36	.04536	22.0444	.04541	22.0217	.00103	.99897	24
37	.04565	21.9041	.04570	21.8813	.00104	.99896	23
38	.04594	.7656	.04599	.7426	.00106	.99894	22
39	.04623	.6288	.04628	.6056	.00107	.99893	21
40	0.04653	21.4937	0.04658	21.4704	1.00108	0.99892	20
41	.04682	.3603	.04687	.3369	.00110	.99890	19
42	.04711	.2285	.04716	.2049	.00111	.99889	18
43	.04740	21.0984	.04745	21.0747	.00113	.99888	17
44	.04769	20.9698	.04774	20.9460	.00114	.99886	16
45	0.04798	20.8428	0.04803	20.8188	1.00115	0.99885	15
46	.04827	.7174	.04833	.6932	.00117	.99883	14
47	.04856	.5934	.04862	.5691	.00118	.99882	13
48	.04885	.4709	.04891	.4465	.00120	.99881	12
49	.04914	.3499	.04920	.3253	.00121	.99879	11
50	0.04943	20.2303	0.04949	20.2056	1.00122	0.99878	10
51	.04972	20.1121	.04978	20.0872	.00124	.99876	9
52	.05001	19.9952	.05007	19.9702	.00125	.99875	8
53	.05030	.8798	.05037	.8546	.00127	.99873	7
54	.05059	.7656	.05066	.7403	.00128	.99872	6
55	0.05088	19.6528	0.05095	19.6273	1.00130	0.99870	5
56	.05117	.5412	.05124	.5156	.00131	.99869	4
57	.05146	.4309	.05153	.4051	.00133	.99867	3
58	.05175	.3218	.05182	.2959	.00134	.99866	2
59	.05205	.2140	.05212	.1879	.00136	.99864	1
60	0.05234	19.1073	0.05241	19.0811	1.00137	0.99863	0
92°→ cos	sec	cot	tan	csc	sin ←87°		

3°→	sin	csc	tan	cot	sec	cos ←176°	
′							′
0	0.05234	19.1073	0.05241	19.0811	1.00137	0.99863	60
1	.05263	19.0019	.05270	18.9755	.00139	.99861	59
2	.05292	18.8975	.05299	.8711	.00140	.99860	58
3	.05321	.7944	.05328	.7678	.00142	.99858	57
4	.05350	.6923	.05357	.6656	.00143	.99857	56
5	0.05379	18.5914	0.05387	18.5645	1.00145	0.99855	55
6	.05408	.4915	.05416	.4645	.00147	.99854	54
7	.05437	.3927	.05445	.3655	.00148	.99852	53
8	.05466	.2950	.05474	.2677	.00150	.99851	52
9	.05495	.1983	.05503	.1708	.00151	.99849	51
10	0.05524	18.1026	0.05533	18.0750	1.00153	0.99847	50
11	.05553	18.0079	.05562	17.9802	.00155	.99846	49
12	.05582	17.9142	.05591	.8863	.00156	.99844	48
13	.05611	.8215	.05620	.7934	.00158	.99842	47
14	.05640	.7298	.05649	.7015	.00159	.99841	46
15	0.05669	17.6389	0.05678	17.6106	1.00161	0.99839	45
16	.05698	.5490	.05708	.5205	.00163	.99838	44
17	.05727	.4600	.05737	.4314	.00164	.99836	43
18	.05756	.3720	.05766	.3432	.00166	.99834	42
19	.05785	.2848	.05795	.2558	.00168	.99833	41
20	0.05814	17.1984	0.05824	17.1693	1.00169	0.99831	40
21	.05844	.1130	.05854	17.0837	.00171	.99829	39
22	.05873	17.0283	.05883	16.9990	.00173	.99827	38
23	.05902	16.9446	.05912	.9150	.00175	.99826	37
24	.05931	.8616	.05941	.8319	.00176	.99824	36
25	0.05960	16.7794	0.05970	16.7496	1.00178	0.99822	35
26	.05989	.6981	.05999	.6681	.00180	.99821	34
27	.06018	.6175	.06029	.5874	.00182	.99819	33
28	.06047	.5377	.06058	.5075	.00183	.99817	32
29	.06076	.4587	.06087	.4283	.00185	.99815	31
30	0.06105	16.3804	0.06116	16.3499	1.00187	0.99813	30
31	.06134	.3029	.06145	.2722	.00189	.99812	29
32	.06163	.2261	.06175	.1952	.00190	.99810	28
33	.06192	.1500	.06204	.1190	.00192	.99808	27
34	.06221	16.0746	.06233	16.0435	.00194	.99806	26
35	0.06250	15.9999	0.06262	15.9687	1.00196	0.99804	25
36	.06279	.9260	.06291	.8945	.00198	.99803	24
37	.06308	.8527	.06321	.8211	.00200	.99801	23
38	.06337	.7801	.06350	.7483	.00201	.99799	22
39	.06366	.7081	.06379	.6762	.00203	.99797	21
40	0.06395	15.6368	0.06408	15.6048	1.00205	0.99795	20
41	.06424	.5661	.06438	.5340	.00207	.99793	19
42	.06453	.4961	.06467	.4638	.00209	.99792	18
43	.06482	.4267	.06496	.3943	.00211	.99790	17
44	.06511	.3579	.06525	.3254	.00213	.99788	16
45	0.06540	15.2898	0.06554	15.2571	1.00215	0.99786	15
46	.06569	.2222	.06584	.1893	.00216	.99784	14
47	.06598	.1553	.06613	.1222	.00218	.99782	13
48	.06627	.0889	.06642	15.0557	.00220	.99780	12
49	.06656	15.0231	.06671	14.9898	.00222	.99778	11
50	0.06685	14.9579	0.06700	14.9244	1.00224	0.99776	10
51	.06714	.8932	.06730	.8596	.00226	.99774	9
52	.06743	.8291	.06759	.7954	.00228	.99772	8
53	.06773	.7656	.06788	.7317	.00230	.99770	7
54	.06802	.7026	.06817	.6685	.00232	.99768	6
55	0.06831	14.6401	0.06847	14.6059	1.00234	0.99766	5
56	.06860	.5782	.06876	.5438	.00236	.99764	4
57	.06889	.5168	.06905	.4823	.00238	.99762	3
58	.06918	.4559	.06934	.4212	.00240	.99760	2
59	.06947	.3955	.06963	.3607	.00242	.99758	1
60	0.06976	14.3356	0.06993	14.3007	1.00244	0.99756	0
93°→ cos	sec	cot	tan	csc	sin ←86°		

4°→	sin	csc	tan	cot	sec	←175° cos	
0	0.06976	14.3356	0.06993	14.3007	1.00244	0.99756	60
1	.07005	.2762	.07022	.2411	.00246	.99754	59
2	.07034	.2173	.07051	.1821	.00248	.99752	58
3	.07063	.1589	.07080	.1235	.00250	.99750	57
4	.07092	.1010	.07110	.0655	.00252	.99748	56
5	0.07121	14.0435	0.07139	14.0079	1.00254	0.99746	55
6	.07150	13.9865	.07168	13.9507	.00257	.99744	54
7	.07179	.9300	.07197	.8940	.00259	.99742	53
8	.07208	.8739	.07227	.8378	.00261	.99740	52
9	.07237	.8183	.07256	.7821	.00263	.99738	51
10	0.07266	13.7631	0.07285	13.7267	1.00265	0.99736	50
11	.07295	.7084	.07314	.6719	.00267	.99734	49
12	.07324	.6541	.07344	.6174	.00269	.99731	48
13	.07353	.6002	.07373	.5634	.00271	.99729	47
14	.07382	.5468	.07402	.5098	.00274	.99727	46
15	0.07411	13.4937	0.07431	13.4566	1.00276	0.99725	45
16	.07440	.4411	.07461	.4039	.00278	.99723	44
17	.07469	.3889	.07490	.3515	.00280	.99721	43
18	.07498	.3371	.07519	.2996	.00282	.99719	42
19	.07527	.2857	.07548	.2480	.00284	.99716	41
20	0.07556	13.2347	0.07578	13.1969	1.00287	0.99714	40
21	.07585	.1841	.07607	.1461	.00289	.99712	39
22	.07614	.1339	.07636	.0958	.00291	.99710	38
23	.07643	.0840	.07665	13.0458	.00293	.99708	37
24	.07672	13.0346	.07695	12.9962	.00296	.99705	36
25	0.07701	12.9855	0.07724	12.9469	1.00298	0.99703	35
26	.07730	.9368	.07753	.8981	.00300	.99701	34
27	.07759	.8884	.07782	.8496	.00302	.99699	33
28	.07788	.8404	.07812	.8014	.00305	.99696	32
29	.07817	.7928	.07841	.7536	.00307	.99694	31
30	0.07846	12.7455	0.07870	12.7062	1.00309	0.99692	30
31	.07875	.6986	.07899	.6591	.00312	.99689	29
32	.07904	.6520	.07929	.6124	.00314	.99687	28
33	.07933	.6057	.07958	.5660	.00316	.99685	27
34	.07962	.5598	.07987	.5199	.00318	.99683	26
35	0.07991	12.5142	0.08017	12.4742	1.00321	0.99680	25
36	.08020	.4690	.08046	.4288	.00323	.99678	24
37	.08049	.4241	.08075	.3838	.00326	.99676	23
38	.08078	.3795	.08104	.3390	.00328	.99673	22
39	.08107	.3352	.08134	.2946	.00330	.99671	21
40	0.08136	12.2913	0.08163	12.2505	1.00333	0.99668	20
41	.08165	.2476	.08192	.2067	.00335	.99666	19
42	.08194	.2043	.08221	.1632	.00337	.99664	18
43	.08223	.1612	.08251	.1201	.00340	.99661	17
44	.08252	.1185	.08280	.0772	.00342	.99659	16
45	0.08281	12.0761	0.08309	12.0346	1.00345	0.99657	15
46	.08310	12.0340	.08339	11.9923	.00347	.99654	14
47	.08339	11.9921	.08368	.9504	.00350	.99652	13
48	.08368	.9506	.08397	.9087	.00352	.99649	12
49	.08397	.9093	.08427	.8673	.00354	.99647	11
50	0.08426	11.8684	0.08456	11.8262	1.00357	0.99644	10
51	.08455	.8277	.08485	.7853	.00359	.99642	9
52	.08484	.7873	.08514	.7448	.00362	.99639	8
53	.08513	.7471	.08544	.7045	.00364	.99637	7
54	.08542	.7073	.08573	.6645	.00367	.99635	6
55	0.08571	11.6677	0.08602	11.6248	1.00369	0.99632	5
56	.08600	.6284	.08632	.5853	.00372	.99630	4
57	.08629	.5893	.08661	.5461	.00374	.99627	3
58	.08658	.5505	.08690	.5072	.00377	.99625	2
59	.08687	.5120	.08720	.4685	.00379	.99622	1
60	0.08716	11.4737	0.08749	11.4301	1.00382	0.99619	0
↑94°→	cos	sec	cot	tan	csc	sin ←85°↑	

5°→	sin	csc	tan	cot	sec	←174° cos	
0	0.08716	11.4737	0.08749	11.4301	1.00382	0.99619	60
1	.08745	.4357	.08778	.3919	.00385	.99617	59
2	.08774	.3979	.08807	.3540	.00387	.99614	58
3	.08803	.3604	.08837	.3163	.00390	.99612	57
4	.08831	.3231	.08866	.2789	.00392	.99609	56
5	0.08860	11.2861	0.08895	11.2417	1.00395	0.99607	55
6	.08889	.2493	.08925	.2048	.00397	.99604	54
7	.08918	.2128	.08954	.1681	.00400	.99602	53
8	.08947	.1765	.08983	.1316	.00403	.99599	52
9	.08976	.1404	.09013	.0954	.00405	.99596	51
10	0.09005	11.1045	0.09042	11.0594	1.00408	0.99594	50
11	.09034	.0689	.09071	11.0237	.00411	.99591	49
12	.09063	11.0336	.09101	10.9882	.00413	.99588	48
13	.09092	10.9984	.09130	.9529	.00416	.99586	47
14	.09121	.9635	.09159	.9178	.00419	.99583	46
15	0.09150	10.9288	0.09189	10.8829	1.00421	0.99580	45
16	.09179	.8943	.09218	.8483	.00424	.99578	44
17	.09208	.8600	.09247	.8139	.00427	.99575	43
18	.09237	.8260	.09277	.7797	.00429	.99572	42
19	.09266	.7921	.09306	.7457	.00432	.99570	41
20	0.09295	10.7585	0.09335	10.7119	1.00435	0.99567	40
21	.09324	.7251	.09365	.6783	.00438	.99564	39
22	.09353	.6919	.09394	.6450	.00440	.99562	38
23	.09382	.6589	.09423	.6118	.00443	.99559	37
24	.09411	.6261	.09453	.5789	.00446	.99556	36
25	0.09440	10.5935	0.09482	10.5462	1.00449	0.99553	35
26	.09469	.5611	.09511	.5136	.00451	.99551	34
27	.09498	.5289	.09541	.4813	.00454	.99548	33
28	.09527	.4969	.09570	.4491	.00457	.99545	32
29	.09556	.4650	.09600	.4172	.00460	.99542	31
30	0.09585	10.4334	0.09629	10.3854	1.00463	0.99540	30
31	.09614	.4020	.09658	.3538	.00465	.99537	29
32	.09642	.3708	.09688	.3224	.00468	.99534	28
33	.09671	.3397	.09717	.2913	.00471	.99531	27
34	.09700	.3089	.09746	.2602	.00474	.99528	26
35	0.09729	10.2782	0.09776	10.2294	1.00477	0.99526	25
36	.09758	.2477	.09805	.1988	.00480	.99523	24
37	.09787	.2174	.09834	.1683	.00482	.99520	23
38	.09816	.1873	.09864	.1381	.00485	.99517	22
39	.09845	.1573	.09893	.1080	.00488	.99514	21
40	0.09874	10.1275	0.09923	10.0780	1.00491	0.99511	20
41	.09903	.0979	.09952	.0483	.00494	.99508	19
42	.09932	.0685	.09981	10.0187	.00497	.99506	18
43	.09961	.0392	.10011	9.98931	.00500	.99503	17
44	.09990	.0101	.10040	.96007	.00503	.99500	16
45	0.10019	9.98123	0.10069	9.93101	1.00506	0.99497	15
46	.10048	.95248	.10099	.90211	.00509	.99494	14
47	.10077	.92389	.10128	.87338	.00512	.99491	13
48	.10106	.89547	.10158	.84482	.00515	.99488	12
49	.10135	.86722	.10187	.81641	.00518	.99485	11
50	0.10164	9.83912	0.10216	9.78817	1.00521	0.99482	10
51	.10192	.81119	.10246	.76009	.00524	.99479	9
52	.10221	.78341	.10275	.73217	.00527	.99476	8
53	.10250	.75579	.10305	.70441	.00530	.99473	7
54	.10279	.72833	.10334	.67680	.00533	.99470	6
55	0.10308	9.70103	0.10363	9.64935	1.00536	0.99467	5
56	.10337	.67387	.10393	.62205	.00539	.99464	4
57	.10366	.64687	.10422	.59490	.00542	.99461	3
58	.10395	.62002	.10452	.56791	.00545	.99458	2
59	.10424	.59332	.10481	.54106	.00548	.99455	1
60	0.10453	9.56677	0.10510	9.51436	1.00551	0.99452	0
↑95°→	cos	sec	cot	tan	csc	sin ←84°↑	

6°→ ↓	sin	csc	tan	cot	sec	cos ←173° ↓
0	0.10453	9.56677	0.10510	9.51436	1.00551	0.99452	60
1	.10482	.54037	.10540	.48781	.00554	.99449	59
2	.10511	.51411	.10569	.46141	.00557	.99446	58
3	.10540	.48800	.10599	.43515	.00560	.99443	57
4	.10569	.46203	.10628	.40904	.00563	.99440	56
5	0.10597	9.43620	0.10657	9.38307	1.00566	0.99437	55
6	.10626	.41052	.10687	.35724	.00569	.99434	54
7	.10655	.38497	.10716	.33155	.00573	.99431	53
8	.10684	.35957	.10746	.30599	.00576	.99428	52
9	.10713	.33430	.10775	.28058	.00579	.99424	51
10	0.10742	9.30917	0.10805	9.25530	1.00582	0.99421	50
11	.10771	.28417	.10834	.23016	.00585	.99418	49
12	.10800	.25931	.10863	.20516	.00588	.99415	48
13	.10829	.23459	.10893	.18028	.00592	.99412	47
14	.10858	.20999	.10922	.15554	.00595	.99409	46
15	0.10887	9.18553	0.10952	9.13093	1.00598	0.99406	45
16	.10916	.16120	.10981	.10646	.00601	.99402	44
17	.10945	.13699	.11011	.08211	.00604	.99399	43
18	.10973	.11292	.11040	.05789	.00608	.99396	42
19	.11002	.08897	.11070	.03379	.00611	.99393	41
20	0.11031	9.06515	0.11099	9.00983	1.00614	0.99390	40
21	.11060	.04146	.11128	8.98598	.00617	.99386	39
22	.11089	9.01788	.11158	.96227	.00621	.99383	38
23	.11118	8.99444	.11187	.93867	.00624	.99380	37
24	.11147	.97111	.11217	.91520	.00627	.99377	36
25	0.11176	8.94791	0.11246	8.89185	1.00630	0.99374	35
26	.11205	.92482	.11276	.86862	.00634	.99370	34
27	.11234	.90186	.11305	.84551	.00637	.99367	33
28	.11263	.87901	.11335	.82252	.00640	.99364	32
29	.11291	.85628	.11364	.79964	.00644	.99360	31
30	0.11320	8.83367	0.11394	8.77689	1.00647	0.99357	30
31	.11349	.81118	.11423	.75425	.00650	.99354	29
32	.11378	.78880	.11452	.73172	.00654	.99351	28
33	.11407	.76653	.11482	.70931	.00657	.99347	27
34	.11436	.74438	.11511	.68701	.00660	.99344	26
35	0.11465	8.72234	0.11541	8.66482	1.00664	0.99341	25
36	.11494	.70041	.11570	.64275	.00667	.99337	24
37	.11523	.67859	.11600	.62078	.00671	.99334	23
38	.11552	.65688	.11629	.59893	.00674	.99331	22
39	.11580	.63528	.11659	.57718	.00677	.99327	21
40	0.11609	8.61379	0.11688	8.55555	1.00681	0.99324	20
41	.11638	.59241	.11718	.53402	.00684	.99320	19
42	.11667	.57113	.11747	.51259	.00688	.99317	18
43	.11696	.54996	.11777	.49126	.00691	.99314	17
44	.11725	.52889	.11806	.47007	.00695	.99310	16
45	0.11754	8.50793	0.11836	8.44896	1.00698	0.99307	15
46	.11783	.48707	.11865	.42795	.00701	.99303	14
47	.11812	.46632	.11895	.40705	.00705	.99300	13
48	.11840	.44566	.11924	.38625	.00708	.99297	12
49	.11869	.42511	.11954	.36555	.00712	.99293	11
50	0.11898	8.40466	0.11983	8.34496	1.00715	0.99290	10
51	.11927	.38431	.12013	.32446	.00719	.99286	9
52	.11956	.36405	.12042	.30406	.00722	.99283	8
53	.11985	.34390	.12072	.28376	.00726	.99279	7
54	.12014	.32384	.12101	.26355	.00730	.99276	6
55	0.12043	8.30388	0.12131	8.24345	1.00733	0.99272	5
56	.12071	.28402	.12160	.22344	.00737	.99269	4
57	.12100	.26425	.12190	.20352	.00740	.99265	3
58	.12129	.24457	.12219	.18370	.00744	.99262	2
59	.12158	.22500	.12249	.16398	.00747	.99258	1
60	0.12187	8.20551	0.12278	8.14435	1.00751	0.99255	0
↑96°→	cos	sec	cot	tan	csc	sin ←83°↑

7°→ ↓	sin	csc	tan	cot	sec	cos ←172° ↓
0	0.12187	8.20551	0.12278	8.14435	1.00751	0.99255	60
1	.12216	.18612	.12308	.12481	.00755	.99251	59
2	.12245	.16681	.12338	.10536	.00758	.99248	58
3	.12274	.14760	.12367	.08600	.00762	.99244	57
4	.12302	.12849	.12397	.06674	.00765	.99240	56
5	0.12331	8.10946	0.12426	8.04756	1.00769	0.99237	55
6	.12360	.09052	.12456	.02848	.00773	.99233	54
7	.12389	.07167	.12485	8.00948	.00776	.99230	53
8	.12418	.05291	.12515	7.99058	.00780	.99226	52
9	.12447	.03423	.12544	.97176	.00784	.99222	51
10	0.12476	8.01565	0.12574	7.95302	1.00787	0.99219	50
11	.12504	7.99714	.12603	.93438	.00791	.99215	49
12	.12533	.97873	.12633	.91582	.00795	.99211	48
13	.12562	.96040	.12662	.89734	.00799	.99208	47
14	.12591	.94216	.12692	.87895	.00802	.99204	46
15	0.12620	7.92399	0.12722	7.86064	1.00806	0.99200	45
16	.12649	.90592	.12751	.84242	.00810	.99197	44
17	.12678	.88792	.12781	.82428	.00813	.99193	43
18	.12706	.87001	.12810	.80622	.00817	.99189	42
19	.12735	.85218	.12840	.78825	.00821	.99186	41
20	0.12764	7.83443	0.12869	7.77035	1.00825	0.99182	40
21	.12793	.81677	.12899	.75254	.00828	.99178	39
22	.12822	.79918	.12929	.73480	.00832	.99175	38
23	.12851	.78167	.12958	.71715	.00836	.99171	37
24	.12880	.76424	.12988	.69957	.00840	.99167	36
25	0.12908	7.74689	0.13017	7.68208	1.00844	0.99163	35
26	.12937	.72962	.13047	.66466	.00848	.99160	34
27	.12966	.71242	.13076	.64732	.00851	.99156	33
28	.12995	.69530	.13106	.63005	.00855	.99152	32
29	.13024	.67826	.13136	.61287	.00859	.99148	31
30	0.13053	7.66130	0.13165	7.59575	1.00863	0.99144	30
31	.13081	.64441	.13195	.57872	.00867	.99141	29
32	.13110	.62759	.13224	.56176	.00871	.99137	28
33	.13139	.61085	.13254	.54487	.00875	.99133	27
34	.13168	.59418	.13284	.52806	.00878	.99129	26
35	0.13197	7.57759	0.13313	7.51132	1.00882	0.99125	25
36	.13226	.56107	.13343	.49465	.00886	.99122	24
37	.13254	.54462	.13372	.47806	.00890	.99118	23
38	.13283	.52825	.13402	.46154	.00894	.99114	22
39	.13312	.51194	.13432	.44509	.00898	.99110	21
40	0.13341	7.49571	0.13461	7.42871	1.00902	0.99106	20
41	.13370	.47955	.13491	.41240	.00906	.99102	19
42	.13399	.46346	.13521	.39616	.00910	.99098	18
43	.13427	.44743	.13550	.37999	.00914	.99094	17
44	.13456	.43148	.13580	.36389	.00918	.99091	16
45	0.13485	7.41560	0.13609	7.34786	1.00922	0.99087	15
46	.13514	.39978	.13639	.33190	.00926	.99083	14
47	.13543	.38403	.13669	.31600	.00930	.99079	13
48	.13572	.36835	.13698	.30018	.00934	.99075	12
49	.13600	.35274	.13728	.28442	.00938	.99071	11
50	0.13629	7.33719	0.13758	7.26873	1.00942	0.99067	10
51	.13658	.32171	.13787	.25310	.00946	.99063	9
52	.13687	.30630	.13817	.23754	.00950	.99059	8
53	.13716	.29095	.13846	.22204	.00954	.99055	7
54	.13744	.27566	.13876	.20661	.00958	.99051	6
55	0.13773	7.26044	0.13906	7.19125	1.00962	0.99047	5
56	.13802	.24529	.13935	.17594	.00966	.99043	4
57	.13831	.23019	.13965	.16071	.00970	.99039	3
58	.13860	.21517	.13995	.14553	.00975	.99035	2
59	.13889	.20020	.14024	.13042	.00979	.99031	1
60	0.13917	7.18530	0.14054	7.11537	1.00983	0.99027	0
↑97°→	cos	sec	cot	tan	csc	sin ←82°↑

330

8°→	sin	csc	tan	cot	sec	cos ←171°	
0	0.13917	7.18530	0.14054	7.11537	1.00983	0.99027	60
1	.13946	.17046	.14084	.10038	.00987	.99023	59
2	.13975	.15568	.14113	.08546	.00991	.99019	58
3	.14004	.14096	.14143	.07059	.00995	.99015	57
4	.14033	.12630	.14173	.05579	.00999	.99011	56
5	0.14061	7.11171	0.14202	7.04105	1.01004	0.99006	55
6	.14090	.09717	.14232	.02637	.01008	.99002	54
7	.14119	.08269	.14262	7.01174	.01012	.98998	53
8	.14148	.06828	.14291	6.99718	.01016	.98994	52
9	.14177	.05392	.14321	.98268	.01020	.98990	51
10	0.14205	7.03962	0.14351	6.96823	1.01024	0.98986	50
11	.14234	.02538	.14381	.95385	.01029	.98982	49
12	.14263	7.01120	.14410	.93952	.01033	.98978	48
13	.14292	6.99708	.14440	.92525	.01037	.98973	47
14	.14320	.98301	.14470	.91104	.01041	.98969	46
15	0.14349	6.96900	0.14499	6.89688	1.01046	0.98965	45
16	.14378	.95505	.14529	.88278	.01050	.98961	44
17	.14407	.94115	.14559	.86874	.01054	.98957	43
18	.14436	.92731	.14588	.85475	.01059	.98953	42
19	.14464	.91352	.14618	.84082	.01063	.98948	41
20	0.14493	6.89979	0.14648	6.82694	1.01067	0.98944	40
21	.14522	.88612	.14678	.81312	.01071	.98940	39
22	.14551	.87250	.14707	.79936	.01076	.98936	38
23	.14580	.85893	.14737	.78564	.01080	.98931	37
24	.14608	.84542	.14767	.77199	.01084	.98927	36
25	0.14637	6.83196	0.14796	6.75838	1.01089	0.98923	35
26	.14666	.81856	.14826	.74483	.01093	.98919	34
27	.14695	.80521	.14856	.73133	.01097	.98914	33
28	.14723	.79191	.14886	.71789	.01102	.98910	32
29	.14752	.77866	.14915	.70450	.01106	.98906	31
30	0.14781	6.76547	0.14945	6.69116	1.01111	0.98902	30
31	.14810	.75233	.14975	.67787	.01115	.98897	29
32	.14838	.73924	.15005	.66463	.01119	.98893	28
33	.14867	.72620	.15034	.65144	.01124	.98889	27
34	.14896	.71321	.15064	.63831	.01128	.98884	26
35	0.14925	6.70027	0.15094	6.62523	1.01133	0.98880	25
36	.14954	.68738	.15124	.61219	.01137	.98876	24
37	.14982	.67454	.15153	.59921	.01142	.98871	23
38	.15011	.66176	.15183	.58627	.01146	.98867	22
39	.15040	.64902	.15213	.57339	.01151	.98863	21
40	0.15069	6.63633	0.15243	6.56055	1.01155	0.98858	20
41	.15097	.62369	.15272	.54777	.01160	.98854	19
42	.15126	.61110	.15302	.53503	.01164	.98849	18
43	.15155	.59855	.15332	.52234	.01169	.98845	17
44	.15184	.58606	.15362	.50970	.01173	.98841	16
45	0.15212	6.57361	0.15391	6.49710	1.01178	0.98836	15
46	.15241	.56121	.15421	.48456	.01182	.98832	14
47	.15270	.54886	.15451	.47206	.01187	.98827	13
48	.15299	.53655	.15481	.45961	.01191	.98823	12
49	.15327	.52429	.15511	.44720	.01196	.98818	11
50	0.15356	6.51208	0.15540	6.43484	1.01200	0.98814	10
51	.15385	.49991	.15570	.42253	.01205	.98809	9
52	.15414	.48779	.15600	.41026	.01209	.98805	8
53	.15442	.47572	.15630	.39804	.01214	.98800	7
54	.15471	.46369	.15660	.38587	.01219	.98796	6
55	0.15500	6.45171	0.15689	6.37374	1.01223	0.98791	5
56	.15529	.43977	.15719	.36165	.01228	.98787	4
57	.15557	.42787	.15749	.34961	.01233	.98782	3
58	.15586	.41602	.15779	.33761	.01237	.98778	2
59	.15615	.40422	.15809	.32566	.01242	.98773	1
60	0.15643	6.39245	0.15838	6.31375	1.01247	0.98769	0
98°→ cos	sec	cot	tan	csc	sin ←81°		

9°→	sin	csc	tan	cot	sec	cos ←170°	
0	0.15643	6.39245	0.15838	6.31375	1.01247	0.98769	60
1	.15672	.38073	.15868	.30189	.01251	.98764	59
2	.15701	.36906	.15898	.29007	.01256	.98760	58
3	.15730	.35743	.15928	.27829	.01261	.98755	57
4	.15758	.34584	.15958	.26655	.01265	.98751	56
5	0.15787	6.33429	0.15988	6.25486	1.01270	0.98746	55
6	.15816	.32279	.16017	.24321	.01275	.98741	54
7	.15845	.31133	.16047	.23160	.01279	.98737	53
8	.15873	.29991	.16077	.22003	.01284	.98732	52
9	.15902	.28853	.16107	.20851	.01289	.98728	51
10	0.15931	6.27719	0.16137	6.19703	1.01294	0.98723	50
11	.15959	.26590	.16167	.18559	.01298	.98718	49
12	.15988	.25464	.16196	.17419	.01303	.98714	48
13	.16017	.24343	.16226	.16283	.01308	.98709	47
14	.16046	.23226	.16256	.15151	.01313	.98704	46
15	0.16074	6.22113	0.16286	6.14023	1.01317	0.98700	45
16	.16103	.21004	.16316	.12899	.01322	.98695	44
17	.16132	.19898	.16346	.11779	.01327	.98690	43
18	.16160	.18797	.16376	.10664	.01332	.98686	42
19	.16189	.17700	.16405	.09552	.01337	.98681	41
20	0.16218	6.16607	0.16435	6.08444	1.01342	0.98676	40
21	.16246	.15517	.16465	.07340	.01346	.98671	39
22	.16275	.14432	.16495	.06240	.01351	.98667	38
23	.16304	.13350	.16525	.05143	.01356	.98662	37
24	.16333	.12273	.16555	.04051	.01361	.98657	36
25	0.16361	6.11199	0.16585	6.02962	1.01366	0.98652	35
26	.16390	.10129	.16615	.01878	.01371	.98648	34
27	.16419	.09062	.16645	6.00797	.01376	.98643	33
28	.16447	.08000	.16674	5.99720	.01381	.98638	32
29	.16476	.06941	.16704	.98646	.01386	.98633	31
30	0.16505	6.05886	0.16734	5.97576	1.01391	0.98629	30
31	.16533	.04834	.16764	.96510	.01395	.98624	29
32	.16562	.03787	.16794	.95448	.01400	.98619	28
33	.16591	.02743	.16824	.94390	.01405	.98614	27
34	.16620	.01702	.16854	.93335	.01410	.98609	26
35	0.16648	6.00666	0.16884	5.92283	1.01415	0.98604	25
36	.16677	5.99633	.16914	.91236	.01420	.98600	24
37	.16706	.98603	.16944	.90191	.01425	.98595	23
38	.16734	.97577	.16974	.89151	.01430	.98590	22
39	.16763	.96555	.17004	.88114	.01435	.98585	21
40	0.16792	5.95536	0.17033	5.87080	1.01440	0.98580	20
41	.16820	.94521	.17063	.86051	.01445	.98575	19
42	.16849	.93509	.17093	.85024	.01450	.98570	18
43	.16878	.92501	.17123	.84001	.01455	.98565	17
44	.16906	.91496	.17153	.82982	.01460	.98561	16
45	0.16935	5.90495	0.17183	5.81966	1.01466	0.98556	15
46	.16964	.89497	.17213	.80953	.01471	.98551	14
47	.16992	.88502	.17243	.79944	.01476	.98546	13
48	.17021	.87511	.17273	.78938	.01481	.98541	12
49	.17050	.86524	.17303	.77936	.01486	.98536	11
50	0.17078	5.85539	0.17333	5.76937	1.01491	0.98531	10
51	.17107	.84558	.17363	.75941	.01496	.98526	9
52	.17136	.83581	.17393	.74949	.01501	.98521	8
53	.17164	.82606	.17423	.73960	.01506	.98516	7
54	.17193	.81635	.17453	.72974	.01512	.98511	6
55	0.17222	5.80667	0.17483	5.71992	1.01517	0.98506	5
56	.17250	.79703	.17513	.71013	.01522	.98501	4
57	.17279	.78742	.17543	.70037	.01527	.98496	3
58	.17308	.77783	.17573	.69064	.01532	.98491	2
59	.17336	.76829	.17603	.68094	.01537	.98486	1
60	0.17365	5.75877	0.17633	5.67128	1.01543	0.98481	0
99°→ cos	sec	cot	tan	csc	sin ←80°		

10°→	sin	csc	tan	cot	sec	cos	←169°
′							′
0	0.17365	5.75877	0.17633	5.67128	1.01543	0.98481	60
1	.17393	.74929	.17663	.66165	.01548	.98476	59
2	.17422	.73983	.17693	.65205	.01553	.98471	58
3	.17451	.73041	.17723	.64248	.01558	.98466	57
4	.17479	.72102	.17753	.63295	.01564	.98461	56
5	0.17508	5.71166	0.17783	5.62344	1.01569	0.98455	55
6	.17537	.70234	.17813	.61397	.01574	.98450	54
7	.17565	.69304	.17843	.60452	.01579	.98445	53
8	.17594	.68377	.17873	.59511	.01585	.98440	52
9	.17623	.67454	.17903	.58573	.01590	.98435	51
10	0.17651	5.66533	0.17933	5.57638	1.01595	0.98430	50
11	.17680	.65616	.17963	.56706	.01601	.98425	49
12	.17708	.64701	.17993	.55777	.01606	.98420	48
13	.17737	.63790	.18023	.54851	.01611	.98414	47
14	.17766	.62881	.18053	.53927	.01616	.98409	46
15	0.17794	5.61976	0.18083	5.53007	1.01622	0.98404	45
16	.17823	.61073	.18113	.52090	.01627	.98399	44
17	.17852	.60174	.18143	.51176	.01633	.98394	43
18	.17880	.59277	.18173	.50264	.01638	.98389	42
19	.17909	.58383	.18203	.49356	.01643	.98383	41
20	0.17937	5.57493	0.18233	5.48451	1.01649	0.98378	40
21	.17966	.56605	.18263	.47548	.01654	.98373	39
22	.17995	.55720	.18293	.46648	.01659	.98368	38
23	.18023	.54837	.18323	.45751	.01665	.98362	37
24	.18052	.53958	.18353	.44857	.01670	.98357	36
25	0.18081	5.53081	0.18384	5.43966	1.01676	0.98352	35
26	.18109	.52208	.18414	.43078	.01681	.98347	34
27	.18138	.51337	.18444	.42192	.01687	.98341	33
28	.18166	.50468	.18474	.41309	.01692	.98336	32
29	.18195	.49603	.18504	.40429	.01698	.98331	31
30	0.18224	5.48740	0.18534	5.39552	1.01703	0.98325	30
31	.18252	.47881	.18564	.38677	.01709	.98320	29
32	.18281	.47023	.18594	.37805	.01714	.98315	28
33	.18309	.46169	.18624	.36936	.01720	.98310	27
34	.18338	.45317	.18654	.36070	.01725	.98304	26
35	0.18367	5.44468	0.18684	5.35206	1.01731	0.98299	25
36	.18395	.43622	.18714	.34345	.01736	.98294	24
37	.18424	.42778	.18745	.33487	.01742	.98288	23
38	.18452	.41937	.18775	.32631	.01747	.98283	22
39	.18481	.41099	.18805	.31778	.01753	.98277	21
40	0.18509	5.40263	0.18835	5.30928	1.01758	0.98272	20
41	.18538	.39430	.18865	.30080	.01764	.98267	19
42	.18567	.38600	.18895	.29235	.01769	.98261	18
43	.18595	.37772	.18925	.28393	.01775	.98256	17
44	.18624	.36947	.18955	.27553	.01781	.98250	16
45	0.18652	5.36124	0.18986	5.26715	1.01786	0.98245	15
46	.18681	.35304	.19016	.25880	.01792	.98240	14
47	.18710	.34486	.19046	.25048	.01798	.98234	13
48	.18738	.33671	.19076	.24218	.01803	.98229	12
49	.18767	.32859	.19106	.23391	.01809	.98223	11
50	0.18795	5.32049	0.19136	5.22566	1.01815	0.98218	10
51	.18824	.31241	.19166	.21744	.01820	.98212	9
52	.18852	.30436	.19197	.20925	.01826	.98207	8
53	.18881	.29634	.19227	.20107	.01832	.98201	7
54	.18910	.28833	.19257	.19293	.01837	.98196	6
55	0.18938	5.28036	0.19287	5.18480	1.01843	0.98190	5
56	.18967	.27241	.19317	.17671	.01849	.98185	4
57	.18995	.26448	.19347	.16863	.01854	.98179	3
58	.19024	.25658	.19378	.16058	.01860	.98174	2
59	.19052	.24870	.19408	.15256	.01866	.98168	1
60	0.19081	5.24084	0.19438	5.14455	1.01872	0.98163	0
100°→	cos	sec	cot	tan	csc	sin	←79°

11°→	sin	csc	tan	cot	sec	cos	←168°
′							′
0	0.19081	5.24084	0.19438	5.14455	1.01872	0.98163	60
1	.19109	.23301	.19468	.13658	.01877	.98157	59
2	.19138	.22521	.19498	.12862	.01883	.98152	58
3	.19167	.21742	.19529	.12069	.01889	.98146	57
4	.19195	.20966	.19559	.11279	.01895	.98140	56
5	0.19224	5.20193	0.19589	5.10490	1.01901	0.98135	55
6	.19252	.19421	.19619	.09704	.01906	.98129	54
7	.19281	.18652	.19649	.08921	.01912	.98124	53
8	.19309	.17886	.19680	.08139	.01918	.98118	52
9	.19338	.17121	.19710	.07360	.01924	.98112	51
10	0.19366	5.16359	0.19740	5.06584	1.01930	0.98107	50
11	.19395	.15599	.19770	.05809	.01936	.98101	49
12	.19423	.14842	.19801	.05037	.01941	.98096	48
13	.19452	.14087	.19831	.04267	.01947	.98090	47
14	.19481	.13334	.19861	.03499	.01953	.98084	46
15	0.19509	5.12583	0.19891	5.02734	1.01959	0.98079	45
16	.19538	.11835	.19921	.01971	.01965	.98073	44
17	.19566	.11088	.19952	.01210	.01971	.98067	43
18	.19595	.10344	.19982	5.00451	.01977	.98061	42
19	.19623	.09602	.20012	4.99695	.01983	.98056	41
20	0.19652	5.08863	0.20042	4.98940	1.01989	0.98050	40
21	.19680	.08125	.20073	.98188	.01995	.98044	39
22	.19709	.07390	.20103	.97438	.02001	.98039	38
23	.19737	.06657	.20133	.96690	.02007	.98033	37
24	.19766	.05926	.20164	.95945	.02013	.98027	36
25	0.19794	5.05197	0.20194	4.95201	1.02019	0.98021	35
26	.19823	.04471	.20224	.94460	.02025	.98016	34
27	.19851	.03746	.20254	.93721	.02031	.98010	33
28	.19880	.03024	.20285	.92984	.02037	.98004	32
29	.19908	.02303	.20315	.92249	.02043	.97998	31
30	0.19937	5.01585	0.20345	4.91516	1.02049	0.97992	30
31	.19965	.00869	.20376	.90785	.02055	.97987	29
32	.19994	5.00155	.20406	.90056	.02061	.97981	28
33	.20022	4.99443	.20436	.89330	.02067	.97975	27
34	.20051	.98733	.20466	.88605	.02073	.97969	26
35	0.20079	4.98025	0.20497	4.87882	1.02079	0.97963	25
36	.20108	.97320	.20527	.87162	.02085	.97958	24
37	.20136	.96616	.20557	.86444	.02091	.97952	23
38	.20165	.95914	.20588	.85727	.02097	.97946	22
39	.20193	.95215	.20618	.85013	.02103	.97940	21
40	0.20222	4.94517	0.20648	4.84300	1.02110	0.97934	20
41	.20250	.93821	.20679	.83590	.02116	.97928	19
42	.20279	.93128	.20709	.82882	.02122	.97922	18
43	.20307	.92436	.20739	.82175	.02128	.97916	17
44	.20336	.91746	.20770	.81471	.02134	.97910	16
45	0.20364	4.91058	0.20800	4.80769	1.02140	0.97905	15
46	.20393	.90373	.20830	.80068	.02146	.97899	14
47	.20421	.89689	.20861	.79370	.02153	.97893	13
48	.20450	.89007	.20891	.78673	.02159	.97887	12
49	.20478	.88327	.20921	.77978	.02165	.97881	11
50	0.20507	4.87649	0.20952	4.77286	1.02171	0.97875	10
51	.20535	.86973	.20982	.76595	.02178	.97869	9
52	.20563	.86299	.21013	.75906	.02184	.97863	8
53	.20592	.85627	.21043	.75219	.02190	.97857	7
54	.20620	.84956	.21073	.74534	.02196	.97851	6
55	0.20649	4.84288	0.21104	4.73851	1.02203	0.97845	5
56	.20677	.83621	.21134	.73170	.02209	.97839	4
57	.20706	.82956	.21164	.72490	.02215	.97833	3
58	.20734	.82294	.21195	.71813	.02221	.97827	2
59	.20763	.81633	.21225	.71137	.02228	.97821	1
60	0.20791	4.80973	0.21256	4.70463	1.02234	0.97815	0
101°→	cos	sec	cot	tan	csc	sin	←78°

12°↓	sin	csc	tan	cot	sec	cos	←167°↓
0	0.20791	4.80973	0.21256	4.70463	1.02234	0.97815	60
1	.20820	.80316	.21286	.69791	.02240	.97809	59
2	.20848	.79661	.21316	.69121	.02247	.97803	58
3	.20877	.79007	.21347	.68452	.02253	.97797	57
4	.20905	.78355	.21377	.67786	.02259	.97791	56
5	0.20933	4.77705	0.21408	4.67121	1.02266	0.97784	55
6	.20962	.77057	.21438	.66458	.02272	.97778	54
7	.20990	.76411	.21469	.65797	.02279	.97772	53
8	.21019	.75766	.21499	.65138	.02285	.97766	52
9	.21047	.75123	.21529	.64480	.02291	.97760	51
10	0.21076	4.74482	0.21560	4.63825	1.02298	0.97754	50
11	.21104	.73843	.21590	.63171	.02304	.97748	49
12	.21132	.73205	.21621	.62518	.02311	.97742	48
13	.21161	.72569	.21651	.61868	.02317	.97735	47
14	.21189	.71935	.21682	.61219	.02323	.97729	46
15	0.21218	4.71303	0.21712	4.60572	1.02330	0.97723	45
16	.21246	.70673	.21743	.59927	.02336	.97717	44
17	.21275	.70044	.21773	.59283	.02343	.97711	43
18	.21303	.69417	.21804	.58641	.02349	.97705	42
19	.21331	.68791	.21834	.58001	.02356	.97698	41
20	0.21360	4.68167	0.21864	4.57363	1.02362	0.97692	40
21	.21388	.67545	.21895	.56726	.02369	.97686	39
22	.21417	.66925	.21925	.56091	.02375	.97680	38
23	.21445	.66307	.21956	.55458	.02382	.97673	37
24	.21474	.65690	.21986	.54826	.02388	.97667	36
25	0.21502	4.65074	0.22017	4.54196	1.02395	0.97661	35
26	.21530	.64461	.22047	.53568	.02402	.97655	34
27	.21559	.63849	.22078	.52941	.02408	.97648	33
28	.21587	.63238	.22108	.52316	.02415	.97642	32
29	.21616	.62630	.22139	.51693	.02421	.97636	31
30	0.21644	4.62023	0.22169	4.51071	1.02428	0.97630	30
31	.21672	.61417	.22200	.50451	.02435	.97623	29
32	.21701	.60813	.22231	.49832	.02441	.97617	28
33	.21729	.60211	.22261	.49215	.02448	.97611	27
34	.21758	.59611	.22292	.48600	.02454	.97604	26
35	0.21786	4.59012	0.22322	4.47986	1.02461	0.97598	25
36	.21814	.58414	.22353	.47374	.02468	.97592	24
37	.21843	.57819	.22383	.46764	.02474	.97585	23
38	.21871	.57224	.22414	.46155	.02481	.97579	22
39	.21899	.56632	.22444	.45548	.02488	.97573	21
40	0.21928	4.56041	0.22475	4.44942	1.02494	0.97566	20
41	.21956	.55451	.22505	.44338	.02501	.97560	19
42	.21985	.54863	.22536	.43735	.02508	.97553	18
43	.22013	.54277	.22567	.43134	.02515	.97547	17
44	.22041	.53692	.22597	.42534	.02521	.97541	16
45	0.22070	4.53109	0.22628	4.41936	1.02528	0.97534	15
46	.22098	.52527	.22658	.41340	.02535	.97528	14
47	.22126	.51947	.22689	.40745	.02542	.97521	13
48	.22155	.51368	.22719	.40152	.02548	.97515	12
49	.22183	.50791	.22750	.39560	.02555	.97508	11
50	0.22212	4.50216	0.22781	4.38969	1.02562	0.97502	10
51	.22240	.49642	.22811	.38381	.02569	.97496	9
52	.22268	.49069	.22842	.37793	.02576	.97489	8
53	.22297	.48498	.22872	.37207	.02582	.97483	7
54	.22325	.47928	.22903	.36623	.02589	.97476	6
55	0.22353	4.47360	0.22934	4.36040	1.02596	0.97470	5
56	.22382	.46793	.22964	.35459	.02603	.97463	4
57	.22410	.46228	.22995	.34879	.02610	.97457	3
58	.22438	.45664	.23026	.34300	.02617	.97450	2
59	.22467	.45102	.23056	.33723	.02624	.97444	1
60	0.22495	4.44541	0.23087	4.33148	1.02630	0.97437	0
102°↑	cos	sec	cot	tan	csc	sin	←77°↑

13°↓	sin	csc	tan	cot	sec	cos	←166°↓
0	0.22495	4.44541	0.23087	4.33148	1.02630	0.97437	60
1	.22523	.43982	.23117	.32573	.02637	.97430	59
2	.22552	.43424	.23148	.32001	.02644	.97424	58
3	.22580	.42867	.23179	.31430	.02651	.97417	57
4	.22608	.42312	.23209	.30860	.02658	.97411	56
5	0.22637	4.41759	0.23240	4.30291	1.02665	0.97404	55
6	.22665	.41206	.23271	.29724	.02672	.97398	54
7	.22693	.40656	.23301	.29159	.02679	.97391	53
8	.22722	.40106	.23332	.28595	.02686	.97384	52
9	.22750	.39558	.23363	.28032	.02693	.97378	51
10	0.22778	4.39012	0.23393	4.27471	1.02700	0.97371	50
11	.22807	.38466	.23424	.26911	.02707	.97365	49
12	.22835	.37923	.23455	.26352	.02714	.97358	48
13	.22863	.37380	.23485	.25795	.02721	.97351	47
14	.22892	.36839	.23516	.25239	.02728	.97345	46
15	0.22920	4.36299	0.23547	4.24685	1.02735	0.97338	45
16	.22948	.35761	.23578	.24132	.02742	.97331	44
17	.22977	.35224	.23608	.23580	.02749	.97325	43
18	.23005	.34689	.23639	.23030	.02756	.97318	42
19	.23033	.34154	.23670	.22481	.02763	.97311	41
20	0.23062	4.33622	0.23700	4.21933	1.02770	0.97304	40
21	.23090	.33090	.23731	.21387	.02777	.97298	39
22	.23118	.32560	.23762	.20842	.02784	.97291	38
23	.23146	.32031	.23793	.20298	.02791	.97284	37
24	.23175	.31503	.23823	.19756	.02799	.97278	36
25	0.23203	4.30977	0.23854	4.19215	1.02806	0.97271	35
26	.23231	.30452	.23885	.18675	.02813	.97264	34
27	.23260	.29929	.23916	.18137	.02820	.97257	33
28	.23288	.29406	.23946	.17600	.02827	.97251	32
29	.23316	.28885	.23977	.17064	.02834	.97244	31
30	0.23345	4.28366	0.24008	4.16530	1.02842	0.97237	30
31	.23373	.27847	.24039	.15997	.02849	.97230	29
32	.23401	.27330	.24069	.15465	.02856	.97223	28
33	.23429	.26814	.24100	.14934	.02863	.97217	27
34	.23458	.26300	.24131	.14405	.02870	.97210	26
35	0.23486	4.25787	0.24162	4.13877	1.02878	0.97203	25
36	.23514	.25275	.24193	.13350	.02885	.97196	24
37	.23542	.24764	.24223	.12825	.02892	.97189	23
38	.23571	.24255	.24254	.12301	.02899	.97182	22
39	.23599	.23746	.24285	.11778	.02907	.97176	21
40	0.23627	4.23239	0.24316	4.11256	1.02914	0.97169	20
41	.23656	.22734	.24347	.10736	.02921	.97162	19
42	.23684	.22229	.24377	.10216	.02928	.97155	18
43	.23712	.21726	.24408	.09699	.02936	.97148	17
44	.23740	.21224	.24439	.09182	.02943	.97141	16
45	0.23769	4.20723	0.24470	4.08666	1.02950	0.97134	15
46	.23797	.20224	.24501	.08152	.02958	.97127	14
47	.23825	.19725	.24532	.07639	.02965	.97120	13
48	.23853	.19228	.24562	.07127	.02972	.97113	12
49	.23882	.18733	.24593	.06616	.02980	.97106	11
50	0.23910	4.18238	0.24624	4.06107	1.02987	0.97100	10
51	.23938	.17744	.24655	.05599	.02994	.97093	9
52	.23966	.17252	.24686	.05092	.03002	.97086	8
53	.23995	.16761	.24717	.04586	.03009	.97079	7
54	.24023	.16271	.24747	.04081	.03017	.97072	6
55	0.24051	4.15782	0.24778	4.03578	1.03024	0.97065	5
56	.24079	.15295	.24809	.03076	.03032	.97058	4
57	.24108	.14809	.24840	.02574	.03039	.97051	3
58	.24136	.14323	.24871	.02074	.03046	.97044	2
59	.24164	.13839	.24902	.01576	.03054	.97037	1
60	0.24192	4.13357	0.24933	4.01078	1.03061	0.97030	0
103°↑	cos	sec	cot	tan	csc	sin	←76°↑

14°→	sin	csc	tan	cot	sec	cos	←165°
0	0.24192	4.13357	0.24933	4.01078	1.03061	0.97030	60
1	.24220	.12875	.24964	.00582	.03069	.97023	59
2	.24249	.12394	.24995	4.00086	.03076	.97015	58
3	.24277	.11915	.25026	3.99592	.03084	.97008	57
4	.24305	.11437	.25056	.99099	.03091	.97001	56
5	0.24333	4.10960	0.25087	3.98607	1.03099	0.96994	55
6	.24362	.10484	.25118	.98117	.03106	.96987	54
7	.24390	.10009	.25149	.97627	.03114	.96980	53
8	.24418	.09535	.25180	.97139	.03121	.96973	52
9	.24446	.09063	.25211	.96651	.03129	.96966	51
10	0.24474	4.08591	0.25242	3.96165	1.03137	0.96959	50
11	.24503	.08121	.25273	.95680	.03144	.96952	49
12	.24531	.07652	.25304	.95196	.03152	.96945	48
13	.24559	.07184	.25335	.94713	.03159	.96937	47
14	.24587	.06717	.25366	.94232	.03167	.96930	46
15	0.24615	4.06251	0.25397	3.93751	1.03175	0.96923	45
16	.24644	.05786	.25428	.93271	.03182	.96916	44
17	.24672	.05322	.25459	.92793	.03190	.96909	43
18	.24700	.04860	.25490	.92316	.03197	.96902	42
19	.24728	.04398	.25521	.91839	.03205	.96894	41
20	0.24756	4.03938	0.25552	3.91364	1.03213	0.96887	40
21	.24784	.03479	.25583	.90890	.03220	.96880	39
22	.24813	.03020	.25614	.90417	.03228	.96873	38
23	.24841	.02563	.25645	.89945	.03236	.96866	37
24	.24869	.02107	.25676	.89474	.03244	.96858	36
25	0.24897	4.01652	0.25707	3.89004	1.03251	0.96851	35
26	.24925	.01198	.25738	.88536	.03259	.96844	34
27	.24954	.00745	.25769	.88068	.03267	.96837	33
28	.24982	4.00293	.25800	.87601	.03275	.96829	32
29	.25010	3.99843	.25831	.87136	.03282	.96822	31
30	0.25038	3.99393	0.25862	3.86671	1.03290	0.96815	30
31	.25066	.98944	.25893	.86208	.03298	.96807	29
32	.25094	.98497	.25924	.85745	.03306	.96800	28
33	.25122	.98050	.25955	.85284	.03313	.96793	27
34	.25151	.97604	.25986	.84824	.03321	.96786	26
35	0.25179	3.97160	0.26017	3.84364	1.03329	0.96778	25
36	.25207	.96716	.26048	.83906	.03337	.96771	24
37	.25235	.96274	.26079	.83449	.03345	.96764	23
38	.25263	.95832	.26110	.82992	.03353	.96756	22
39	.25291	.95392	.26141	.82537	.03360	.96749	21
40	0.25320	3.94952	0.26172	3.82083	1.03368	0.96742	20
41	.25348	.94514	.26203	.81630	.03376	.96734	19
42	.25376	.94076	.26235	.81177	.03384	.96727	18
43	.25404	.93640	.26266	.80726	.03392	.96719	17
44	.25432	.93204	.26297	.80276	.03400	.96712	16
45	0.25460	3.92770	0.26328	3.79827	1.03408	0.96705	15
46	.25488	.92337	.26359	.79378	.03416	.96697	14
47	.25516	.91904	.26390	.78931	.03424	.96690	13
48	.25545	.91473	.26421	.78485	.03432	.96682	12
49	.25573	.91042	.26452	.78040	.03439	.96675	11
50	0.25601	3.90613	0.26483	3.77595	1.03447	0.96667	10
51	.25629	.90184	.26515	.77152	.03455	.96660	9
52	.25657	.89756	.26546	.76709	.03463	.96653	8
53	.25685	.89330	.26577	.76268	.03471	.96645	7
54	.25713	.88904	.26608	.75828	.03479	.96638	6
55	0.25741	3.88479	0.26639	3.75388	1.03487	0.96630	5
56	.25769	.88056	.26670	.74950	.03495	.96623	4
57	.25798	.87633	.26701	.74512	.03503	.96615	3
58	.25826	.87211	.26733	.74075	.03511	.96608	2
59	.25854	.86790	.26764	.73640	.03520	.96600	1
60	0.25882	3.86370	0.26795	3.73205	1.03528	0.96593	0
104°→ cos	sec	cot	tan	csc	sin ←75°		

15°→	sin	csc	tan	cot	sec	cos	←164°
0	0.25882	3.86370	0.26795	3.73205	1.03528	0.96593	60
1	.25910	.85951	.26826	.72771	.03536	.96585	59
2	.25938	.85533	.26857	.72338	.03544	.96578	58
3	.25966	.85116	.26888	.71907	.03552	.96570	57
4	.25994	.84700	.26920	.71476	.03560	.96562	56
5	0.26022	3.84285	0.26951	3.71046	1.03568	0.96555	55
6	.26050	.83871	.26982	.70616	.03576	.96547	54
7	.26079	.83457	.27013	.70188	.03584	.96540	53
8	.26107	.83045	.27044	.69761	.03592	.96532	52
9	.26135	.82633	.27076	.69335	.03601	.96524	51
10	0.26163	3.82223	0.27107	3.68909	1.03609	0.96517	50
11	.26191	.81813	.27138	.68485	.03617	.96509	49
12	.26219	.81404	.27169	.68061	.03625	.96502	48
13	.26247	.80996	.27201	.67638	.03633	.96494	47
14	.26275	.80589	.27232	.67217	.03642	.96486	46
15	0.26303	3.80183	0.27263	3.66796	1.03650	0.96479	45
16	.26331	.79778	.27294	.66376	.03658	.96471	44
17	.26359	.79374	.27326	.65957	.03666	.96463	43
18	.26387	.78970	.27357	.65538	.03674	.96456	42
19	.26415	.78568	.27388	.65121	.03683	.96448	41
20	0.26443	3.78166	0.27419	3.64705	1.03691	0.96440	40
21	.26471	.77765	.27451	.64289	.03699	.96433	39
22	.26500	.77365	.27482	.63874	.03708	.96425	38
23	.26528	.76966	.27513	.63461	.03716	.96417	37
24	.26556	.76568	.27545	.63048	.03724	.96410	36
25	0.26584	3.76171	0.27576	3.62636	1.03732	0.96402	35
26	.26612	.75775	.27607	.62224	.03741	.96394	34
27	.26640	.75379	.27638	.61814	.03749	.96386	33
28	.26668	.74984	.27670	.61405	.03757	.96379	32
29	.26696	.74591	.27701	.60996	.03766	.96371	31
30	0.26724	3.74198	0.27732	3.60588	1.03774	0.96363	30
31	.26752	.73806	.27764	.60181	.03783	.96355	29
32	.26780	.73414	.27795	.59775	.03791	.96347	28
33	.26808	.73024	.27826	.59370	.03799	.96340	27
34	.26836	.72635	.27858	.58966	.03808	.96332	26
35	0.26864	3.72246	0.27889	3.58562	1.03816	0.96324	25
36	.26892	.71858	.27921	.58160	.03825	.96316	24
37	.26920	.71471	.27952	.57758	.03833	.96308	23
38	.26948	.71085	.27983	.57357	.03842	.96301	22
39	.26976	.70700	.28015	.56957	.03850	.96293	21
40	0.27004	3.70315	0.28046	3.56557	1.03858	0.96285	20
41	.27032	.69931	.28077	.56159	.03867	.96277	19
42	.27060	.69549	.28109	.55761	.03875	.96269	18
43	.27088	.69167	.28140	.55364	.03884	.96261	17
44	.27116	.68785	.28172	.54968	.03892	.96253	16
45	0.27144	3.68405	0.28203	3.54573	1.03901	0.96246	15
46	.27172	.68025	.28234	.54179	.03909	.96238	14
47	.27200	.67647	.28266	.53785	.03918	.96230	13
48	.27228	.67269	.28297	.53393	.03927	.96222	12
49	.27256	.66892	.28329	.53001	.03935	.96214	11
50	0.27284	3.66515	0.28360	3.52609	1.03944	0.96206	10
51	.27312	.66140	.28391	.52219	.03952	.96198	9
52	.27340	.65765	.28423	.51829	.03961	.96190	8
53	.27368	.65391	.28454	.51441	.03969	.96182	7
54	.27396	.65018	.28486	.51053	.03978	.96174	6
55	0.27424	3.64645	0.28517	3.50666	1.03987	0.96166	5
56	.27452	.64274	.28549	.50279	.03995	.96158	4
57	.27480	.63903	.28580	.49894	.04004	.96150	3
58	.27508	.63533	.28612	.49509	.04013	.96142	2
59	.27536	.63164	.28643	.49125	.04021	.96134	1
60	0.27564	3.62796	0.28675	3.48741	1.04030	0.96126	0
105°→ cos	sec	cot	tan	csc	sin ←74°		

16° ↓	sin	csc	tan	cot	sec	cos ←163° ↓	
0	0.27564	3.62796	0.28675	3.48741	1.04030	0.96126	60
1	.27592	.62428	.28706	.48359	.04039	.96118	59
2	.27620	.62061	.28738	.47977	.04047	.96110	58
3	.27648	.61695	.28769	.47596	.04056	.96102	57
4	.27676	.61330	.28801	.47216	.04065	.96094	56
5	0.27704	3.60965	0.28832	3.46837	1.04073	0.96086	55
6	.27731	.60601	.28864	.46458	.04082	.96078	54
7	.27759	.60238	.28895	.46080	.04091	.96070	53
8	.27787	.59876	.28927	.45703	.04100	.96062	52
9	.27815	.59514	.28958	.45327	.04108	.96054	51
10	0.27843	3.59154	0.28990	3.44951	1.04117	0.96046	50
11	.27871	.58794	.29021	.44576	.04126	.96037	49
12	.27899	.58434	.29053	.44202	.04135	.96029	48
13	.27927	.58076	.29084	.43829	.04144	.96021	47
14	.27955	.57718	.29116	.43456	.04152	.96013	46
15	0.27983	3.57361	0.29147	3.43084	1.04161	0.96005	45
16	.28011	.57005	.29179	.42713	.04170	.95997	44
17	.28039	.56649	.29210	.42343	.04179	.95989	43
18	.28067	.56294	.29242	.41973	.04188	.95981	42
19	.28095	.55940	.29274	.41604	.04197	.95972	41
20	0.28123	3.55587	0.29305	3.41236	1.04206	0.95964	40
21	.28150	.55234	.29337	.40869	.04214	.95956	39
22	.28178	.54883	.29368	.40502	.04223	.95948	38
23	.28206	.54531	.29400	.40136	.04232	.95940	37
24	.28234	.54181	.29432	.39771	.04241	.95931	36
25	0.28262	3.53831	0.29463	3.39406	1.04250	0.95923	35
26	.28290	.53482	.29495	.39042	.04259	.95915	34
27	.28318	.53134	.29526	.38679	.04268	.95907	33
28	.28346	.52787	.29558	.38317	.04277	.95898	32
29	.28374	.52440	.29590	.37955	.04286	.95890	31
30	0.28402	3.52094	0.29621	3.37594	1.04295	0.95882	30
31	.28429	.51748	.29653	.37234	.04304	.95874	29
32	.28457	.51404	.29685	.36875	.04313	.95865	28
33	.28485	.51060	.29716	.36516	.04322	.95857	27
34	.28513	.50716	.29748	.36158	.04331	.95849	26
35	0.28541	3.50374	0.29780	3.35800	1.04340	0.95841	25
36	.28569	.50032	.29811	.35443	.04349	.95832	24
37	.28597	.49691	.29843	.35087	.04358	.95824	23
38	.28625	.49350	.29875	.34732	.04367	.95816	22
39	.28652	.49010	.29906	.34377	.04376	.95807	21
40	0.28680	3.48671	0.29938	3.34023	1.04385	0.95799	20
41	.28708	.48333	.29970	.33670	.04394	.95791	19
42	.28736	.47995	.30001	.33317	.04403	.95782	18
43	.28764	.47658	.30033	.32965	.04413	.95774	17
44	.28792	.47321	.30065	.32614	.04422	.95766	16
45	0.28820	3.46986	0.30097	3.32264	1.04431	0.95757	15
46	.28847	.46651	.30128	.31914	.04440	.95749	14
47	.28875	.46316	.30160	.31565	.04449	.95740	13
48	.28903	.45983	.30192	.31216	.04458	.95732	12
49	.28931	.45650	.30224	.30868	.04468	.95724	11
50	0.28959	3.45317	0.30255	3.30521	1.04477	0.95715	10
51	.28987	.44986	.30287	.30174	.04486	.95707	9
52	.29015	.44655	.30319	.29829	.04495	.95698	8
53	.29042	.44324	.30351	.29483	.04504	.95690	7
54	.29070	.43995	.30382	.29139	.04514	.95681	6
55	0.29098	3.43666	0.30414	3.28795	1.04523	0.95673	5
56	.29126	.43337	.30446	.28452	.04532	.95664	4
57	.29154	.43010	.30478	.28109	.04541	.95656	3
58	.29182	.42683	.30509	.27767	.04551	.95647	2
59	.29209	.42356	.30541	.27426	.04560	.95639	1
60	0.29237	3.42030	0.30573	3.27085	1.04569	0.95630	0
↑ 106°→	cos	sec	cot	tan	csc	sin ←73° ↑	

17° ↓	sin	csc	tan	cot	sec	cos ←162° ↓	
0	0.29237	3.42030	0.30573	3.27085	1.04569	0.95630	60
1	.29265	.41705	.30605	.26745	.04578	.95622	59
2	.29293	.41381	.30637	.26406	.04588	.95613	58
3	.29321	.41057	.30669	.26067	.04597	.95605	57
4	.29348	.40734	.30700	.25729	.04606	.95596	56
5	0.29376	3.40411	0.30732	3.25392	1.04616	0.95588	55
6	.29404	.40089	.30764	.25055	.04625	.95579	54
7	.29432	.39768	.30796	.24719	.04635	.95571	53
8	.29460	.39448	.30828	.24383	.04644	.95562	52
9	.29487	.39128	.30860	.24049	.04653	.95554	51
10	0.29515	3.38808	0.30891	3.23714	1.04663	0.95545	50
11	.29543	.38489	.30923	.23381	.04672	.95536	49
12	.29571	.38171	.30955	.23048	.04682	.95528	48
13	.29599	.37854	.30987	.22715	.04691	.95519	47
14	.29626	.37537	.31019	.22384	.04700	.95511	46
15	0.29654	3.37221	0.31051	3.22053	1.04710	0.95502	45
16	.29682	.36905	.31083	.21722	.04719	.95493	44
17	.29710	.36590	.31115	.21392	.04729	.95485	43
18	.29737	.36276	.31147	.21063	.04738	.95476	42
19	.29765	.35962	.31178	.20734	.04748	.95467	41
20	0.29793	3.35649	0.31210	3.20406	1.04757	0.95459	40
21	.29821	.35336	.31242	.20079	.04767	.95450	39
22	.29849	.35025	.31274	.19752	.04776	.95441	38
23	.29876	.34713	.31306	.19426	.04786	.95433	37
24	.29904	.34403	.31338	.19100	.04795	.95424	36
25	0.29932	3.34092	0.31370	3.18775	1.04805	0.95415	35
26	.29960	.33783	.31402	.18451	.04815	.95407	34
27	.29987	.33474	.31434	.18127	.04824	.95398	33
28	.30015	.33166	.31466	.17804	.04834	.95389	32
29	.30043	.32858	.31498	.17481	.04843	.95380	31
30	0.30071	3.32551	0.31530	3.17159	1.04853	0.95372	30
31	.30098	.32244	.31562	.16838	.04863	.95363	29
32	.30126	.31939	.31594	.16517	.04872	.95354	28
33	.30154	.31633	.31626	.16197	.04882	.95345	27
34	.30182	.31328	.31658	.15877	.04891	.95337	26
35	0.30209	3.31024	0.31690	3.15558	1.04901	0.95328	25
36	.30237	.30721	.31722	.15240	.04911	.95319	24
37	.30265	.30418	.31754	.14922	.04920	.95310	23
38	.30292	.30115	.31786	.14605	.04930	.95301	22
39	.30320	.29814	.31818	.14288	.04940	.95293	21
40	0.30348	3.29512	0.31850	3.13972	1.04950	0.95284	20
41	.30376	.29212	.31882	.13656	.04959	.95275	19
42	.30403	.28912	.31914	.13341	.04969	.95266	18
43	.30431	.28612	.31946	.13027	.04979	.95257	17
44	.30459	.28313	.31978	.12713	.04989	.95248	16
45	0.30486	3.28015	0.32010	3.12400	1.04998	0.95240	15
46	.30514	.27717	.32042	.12087	.05008	.95231	14
47	.30542	.27420	.32074	.11775	.05018	.95222	13
48	.30570	.27123	.32106	.11464	.05028	.95213	12
49	.30597	.26827	.32139	.11153	.05038	.95204	11
50	0.30625	3.26531	0.32171	3.10842	1.05047	0.95195	10
51	.30653	.26237	.32203	.10532	.05057	.95186	9
52	.30680	.25942	.32235	.10223	.05067	.95177	8
53	.30708	.25648	.32267	.09914	.05077	.95168	7
54	.30736	.25355	.32299	.09606	.05087	.95159	6
55	0.30763	3.25062	0.32331	3.09298	1.05097	0.95150	5
56	.30791	.24770	.32363	.08991	.05107	.95142	4
57	.30819	.24478	.32396	.08685	.05116	.95133	3
58	.30846	.24187	.32428	.08379	.05126	.95124	2
59	.30874	.23897	.32460	.08073	.05136	.95115	1
60	0.30902	3.23607	0.32492	3.07768	1.05146	0.95106	0
↑ 107°→	cos	sec	cot	tan	csc	sin ←72° ↑	

18°→	sin	csc	tan	cot	sec	cos	←161°
′							′
0	0.30902	3.23607	0.32492	3.07768	1.05146	0.95106	60
1	.30929	.23317	.32524	.07464	.05156	.95097	59
2	.30957	.23028	.32556	.07160	.05166	.95088	58
3	.30985	.22740	.32588	.06857	.05176	.95079	57
4	.31012	.22452	.32621	.06554	.05186	.95070	56
5	0.31040	3.22165	0.32653	3.06252	1.05196	0.95061	55
6	.31068	.21878	.32685	.05950	.05206	.95052	54
7	.31095	.21592	.32717	.05649	.05216	.95043	53
8	.31123	.21306	.32749	.05349	.05226	.95033	52
9	.31151	.21021	.32782	.05049	.05236	.95024	51
10	0.31178	3.20737	0.32814	3.04749	1.05246	0.95015	50
11	.31206	.20453	.32846	.04450	.05256	.95006	49
12	.31233	.20169	.32878	.04152	.05266	.94997	48
13	.31261	.19886	.32911	.03854	.05276	.94988	47
14	.31289	.19604	.32943	.03556	.05286	.94979	46
15	0.31316	3.19322	0.32975	3.03260	1.05297	0.94970	45
16	.31344	.19040	.33007	.02963	.05307	.94961	44
17	.31372	.18759	.33040	.02667	.05317	.94952	43
18	.31399	.18479	.33072	.02372	.05327	.94943	42
19	.31427	.18199	.33104	.02077	.05337	.94933	41
20	0.31454	3.17920	0.33136	3.01783	1.05347	0.94924	40
21	.31482	.17641	.33169	.01489	.05357	.94915	39
22	.31510	.17363	.33201	.01196	.05367	.94906	38
23	.31537	.17085	.33233	.00903	.05378	.94897	37
24	.31565	.16808	.33266	.00611	.05388	.94888	36
25	0.31593	3.16531	0.33298	3.00319	1.05398	0.94878	35
26	.31620	.16255	.33330	3.00028	.05408	.94869	34
27	.31648	.15979	.33363	2.99738	.05418	.94860	33
28	.31675	.15704	.33395	.99447	.05429	.94851	32
29	.31703	.15429	.33427	.99158	.05439	.94842	31
30	0.31730	3.15155	0.33460	2.98868	1.05449	0.94832	30
31	.31758	.14881	.33492	.98580	.05459	.94823	29
32	.31786	.14608	.33524	.98292	.05470	.94814	28
33	.31813	.14335	.33557	.98004	.05480	.94805	27
34	.31841	.14063	.33589	.97717	.05490	.94795	26
35	0.31868	3.13791	0.33621	2.97430	1.05501	0.94786	25
36	.31896	.13520	.33654	.97144	.05511	.94777	24
37	.31923	.13249	.33686	.96858	.05521	.94768	23
38	.31951	.12979	.33718	.96573	.05532	.94758	22
39	.31979	.12709	.33751	.96288	.05542	.94749	21
40	0.32006	3.12440	0.33783	2.96004	1.05552	0.94740	20
41	.32034	.12171	.33816	.95721	.05563	.94730	19
42	.32061	.11903	.33848	.95437	.05573	.94721	18
43	.32089	.11635	.33881	.95155	.05584	.94712	17
44	.32116	.11367	.33913	.94872	.05594	.94702	16
45	0.32144	3.11101	0.33945	2.94591	1.05604	0.94693	15
46	.32171	.10834	.33978	.94309	.05615	.94684	14
47	.32199	.10568	.34010	.94028	.05625	.94674	13
48	.32227	.10303	.34043	.93748	.05636	.94665	12
49	.32254	.10038	.34075	.93468	.05646	.94656	11
50	0.32282	3.09774	0.34108	2.93189	1.05657	0.94646	10
51	.32309	.09510	.34140	.92910	.05667	.94637	9
52	.32337	.09246	.34173	.92632	.05678	.94627	8
53	.32364	.08983	.34205	.92354	.05688	.94618	7
54	.32392	.08721	.34238	.92076	.05699	.94609	6
55	0.32419	3.08459	0.34270	2.91799	1.05709	0.94599	5
56	.32447	.08197	.34303	.91523	.05720	.94590	4
57	.32474	.07936	.34335	.91246	.05730	.94580	3
58	.32502	.07675	.34368	.90971	.05741	.94571	2
59	.32529	.07415	.34400	.90696	.05751	.94561	1
60	0.32557	3.07155	0.34433	2.90421	1.05762	0.94552	0
108°→ cos	sec	cot	tan	csc	sin ←71°		

19°→	sin	csc	tan	cot	sec	cos	←160°
′							′
0	0.32557	3.07155	0.34433	2.90421	1.05762	0.94552	60
1	.32584	.06896	.34465	.90147	.05773	.94542	59
2	.32612	.06637	.34498	.89873	.05783	.94533	58
3	.32639	.06379	.34530	.89600	.05794	.94523	57
4	.32667	.06121	.34563	.89327	.05805	.94514	56
5	0.32694	3.05864	0.34596	2.89055	1.05815	0.94504	55
6	.32722	.05607	.34628	.88783	.05826	.94495	54
7	.32749	.05350	.34661	.88511	.05836	.94485	53
8	.32777	.05094	.34693	.88240	.05847	.94476	52
9	.32804	.04839	.34726	.87970	.05858	.94466	51
10	0.32832	3.04584	0.34758	2.87700	1.05869	0.94457	50
11	.32859	.04329	.34791	.87430	.05879	.94447	49
12	.32887	.04075	.34824	.87161	.05890	.94438	48
13	.32914	.03821	.34856	.86892	.05901	.94428	47
14	.32942	.03568	.34889	.86624	.05911	.94418	46
15	0.32969	3.03315	0.34922	2.86356	1.05922	0.94409	45
16	.32997	.03062	.34954	.86089	.05933	.94399	44
17	.33024	.02810	.34987	.85822	.05944	.94390	43
18	.33051	.02559	.35020	.85555	.05955	.94380	42
19	.33079	.02308	.35052	.85289	.05965	.94370	41
20	0.33106	3.02057	0.35085	2.85023	1.05976	0.94361	40
21	.33134	.01807	.35118	.84758	.05987	.94351	39
22	.33161	.01557	.35150	.84494	.05998	.94342	38
23	.33189	.01308	.35183	.84229	.06009	.94332	37
24	.33216	.01059	.35216	.83965	.06020	.94322	36
25	0.33244	3.00810	0.35248	2.83702	1.06030	0.94313	35
26	.33271	.00562	.35281	.83439	.06041	.94303	34
27	.33298	.00315	.35314	.83176	.06052	.94293	33
28	.33326	3.00067	.35346	.82914	.06063	.94284	32
29	.33353	2.99821	.35379	.82653	.06074	.94274	31
30	0.33381	2.99574	0.35412	2.82391	1.06085	0.94264	30
31	.33408	.99329	.35445	.82130	.06096	.94254	29
32	.33436	.99083	.35477	.81870	.06107	.94245	28
33	.33463	.98838	.35510	.81610	.06118	.94235	27
34	.33490	.98594	.35543	.81350	.06129	.94225	26
35	0.33518	2.98349	0.35576	2.81091	1.06140	0.94215	25
36	.33545	.98106	.35608	.80833	.06151	.94206	24
37	.33573	.97862	.35641	.80574	.06162	.94196	23
38	.33600	.97619	.35674	.80316	.06173	.94186	22
39	.33627	.97377	.35707	.80059	.06184	.94176	21
40	0.33655	2.97135	0.35740	2.79802	1.06195	0.94167	20
41	.33682	.96893	.35772	.79545	.06206	.94157	19
42	.33710	.96652	.35805	.79289	.06217	.94147	18
43	.33737	.96411	.35838	.79033	.06228	.94137	17
44	.33764	.96171	.35871	.78778	.06239	.94127	16
45	0.33792	2.95931	0.35904	2.78523	1.06250	0.94118	15
46	.33819	.95691	.35937	.78269	.06261	.94108	14
47	.33846	.95452	.35969	.78014	.06272	.94098	13
48	.33874	.95213	.36002	.77761	.06283	.94088	12
49	.33901	.94975	.36035	.77507	.06295	.94078	11
50	0.33929	2.94737	0.36068	2.77254	1.06306	0.94068	10
51	.33956	.94500	.36101	.77002	.06317	.94058	9
52	.33983	.94263	.36134	.76750	.06328	.94049	8
53	.34011	.94026	.36167	.76498	.06339	.94039	7
54	.34038	.93790	.36199	.76247	.06350	.94029	6
55	0.34065	2.93554	0.36232	2.75996	1.06362	0.94019	5
56	.34093	.93318	.36265	.75746	.06373	.94009	4
57	.34120	.93083	.36298	.75496	.06384	.93999	3
58	.34147	.92849	.36331	.75246	.06395	.93989	2
59	.34175	.92614	.36364	.74997	.06407	.93979	1
60	0.34202	2.92380	0.36397	2.74748	1.06418	0.93969	0
109°→ cos	sec	cot	tan	csc	sin ←70°		

20°→	sin	csc	tan	cot	sec	cos	←159°
0	0.34202	2.92380	0.36397	2.74748	1.06418	0.93969	60
1	.34229	.92147	.36430	.74499	.06429	.93959	59
2	.34257	.91914	.36463	.74251	.06440	.93949	58
3	.34284	.91681	.36496	.74004	.06452	.93939	57
4	.34311	.91449	.36529	.73756	.06463	.93929	56
5	0.34339	2.91217	0.36562	2.73509	1.06474	0.93919	55
6	.34366	.90986	.36595	.73263	.06486	.93909	54
7	.34393	.90754	.36628	.73017	.06497	.93899	53
8	.34421	.90524	.36661	.72771	.06508	.93889	52
9	.34448	.90293	.36694	.72526	.06520	.93879	51
10	0.34475	2.90063	0.36727	2.72281	1.06531	0.93869	50
11	.34503	.89834	.36760	.72036	.06542	.93859	49
12	.34530	.89605	.36793	.71792	.06554	.93849	48
13	.34557	.89376	.36826	.71548	.06565	.93839	47
14	.34584	.89148	.36859	.71305	.06577	.93829	46
15	0.34612	2.88920	0.36892	2.71062	1.06588	0.93819	45
16	.34639	.88692	.36925	.70819	.06600	.93809	44
17	.34666	.88465	.36958	.70577	.06611	.93799	43
18	.34694	.88238	.36991	.70335	.06622	.93789	42
19	.34721	.88011	.37024	.70094	.06634	.93779	41
20	0.34748	2.87785	0.37057	2.69853	1.06645	0.93769	40
21	.34775	.87560	.37090	.69612	.06657	.93759	39
22	.34803	.87334	.37123	.69371	.06668	.93748	38
23	.34830	.87109	.37157	.69131	.06680	.93738	37
24	.34857	.86885	.37190	.68892	.06691	.93728	36
25	0.34884	2.86661	0.37223	2.68653	1.06703	0.93718	35
26	.34912	.86437	.37256	.68414	.06715	.93708	34
27	.34939	.86213	.37289	.68175	.06726	.93698	33
28	.34966	.85990	.37322	.67937	.06738	.93688	32
29	.34993	.85767	.37355	.67700	.06749	.93677	31
30	0.35021	2.85545	0.37388	2.67462	1.06761	0.93667	30
31	.35048	.85323	.37422	.67225	.06773	.93657	29
32	.35075	.85102	.37455	.66989	.06784	.93647	28
33	.35102	.84880	.37488	.66752	.06796	.93637	27
34	.35130	.84659	.37521	.66516	.06807	.93626	26
35	0.35157	2.84439	0.37554	2.66281	1.06819	0.93616	25
36	.35184	.84219	.37588	.66046	.06831	.93606	24
37	.35211	.83999	.37621	.65811	.06842	.93596	23
38	.35239	.83780	.37654	.65576	.06854	.93585	22
39	.35266	.83561	.37687	.65342	.06866	.93575	21
40	0.35293	2.83342	0.37720	2.65109	1.06878	0.93565	20
41	.35320	.83124	.37754	.64875	.06889	.93555	19
42	.35347	.82906	.37787	.64642	.06901	.93544	18
43	.35375	.82688	.37820	.64410	.06913	.93534	17
44	.35402	.82471	.37853	.64177	.06925	.93524	16
45	0.35429	2.82254	0.37887	2.63945	1.06936	0.93514	15
46	.35456	.82037	.37920	.63714	.06948	.93503	14
47	.35484	.81821	.37953	.63483	.06960	.93493	13
48	.35511	.81605	.37986	.63252	.06972	.93483	12
49	.35538	.81390	.38020	.63021	.06984	.93472	11
50	0.35565	2.81175	0.38053	2.62791	1.06995	0.93462	10
51	.35592	.80960	.38086	.62561	.07007	.93452	9
52	.35619	.80746	.38120	.62332	.07019	.93441	8
53	.35647	.80531	.38153	.62103	.07031	.93431	7
54	.35674	.80318	.38186	.61874	.07043	.93420	6
55	0.35701	2.80104	0.38220	2.61646	1.07055	0.93410	5
56	.35728	.79891	.38253	.61418	.07067	.93400	4
57	.35755	.79679	.38286	.61190	.07079	.93389	3
58	.35782	.79466	.38320	.60963	.07091	.93379	2
59	.35810	.79254	.38353	.60736	.07103	.93368	1
60	0.35837	2.79043	0.38386	2.60509	1.07114	0.93358	0
110°→	cos	sec	cot	tan	csc	sin	←69°

21°→	sin	csc	tan	cot	sec	cos	←158°
0	0.35837	2.79043	0.38386	2.60509	1.07114	0.93358	60
1	.35864	.78832	.38420	.60283	.07126	.93348	59
2	.35891	.78621	.38453	.60057	.07138	.93337	58
3	.35918	.78410	.38487	.59831	.07150	.93327	57
4	.35945	.78200	.38520	.59606	.07162	.93316	56
5	0.35973	2.77990	0.38553	2.59381	1.07174	0.93306	55
6	.36000	.77780	.38587	.59156	.07186	.93295	54
7	.36027	.77571	.38620	.58932	.07199	.93285	53
8	.36054	.77362	.38654	.58708	.07211	.93274	52
9	.36081	.77154	.38687	.58484	.07223	.93264	51
10	0.36108	2.76945	0.38721	2.58261	1.07235	0.93253	50
11	.36135	.76737	.38754	.58038	.07247	.93243	49
12	.36162	.76530	.38787	.57815	.07259	.93232	48
13	.36190	.76323	.38821	.57593	.07271	.93222	47
14	.36217	.76116	.38854	.57371	.07283	.93211	46
15	0.36244	2.75909	0.38888	2.57150	1.07295	0.93201	45
16	.36271	.75703	.38921	.56928	.07307	.93190	44
17	.36298	.75497	.38955	.56707	.07320	.93180	43
18	.36325	.75292	.38988	.56487	.07332	.93169	42
19	.36352	.75086	.39022	.56266	.07344	.93159	41
20	0.36379	2.74881	0.39055	2.56046	1.07356	0.93148	40
21	.36406	.74677	.39089	.55827	.07368	.93137	39
22	.36434	.74473	.39122	.55608	.07380	.93127	38
23	.36461	.74269	.39156	.55389	.07393	.93116	37
24	.36488	.74065	.39190	.55170	.07405	.93106	36
25	0.36515	2.73862	0.39223	2.54952	1.07417	0.93095	35
26	.36542	.73659	.39257	.54734	.07429	.93084	34
27	.36569	.73456	.39290	.54516	.07442	.93074	33
28	.36596	.73254	.39324	.54299	.07454	.93063	32
29	.36623	.73052	.39357	.54082	.07466	.93052	31
30	0.36650	2.72850	0.39391	2.53865	1.07479	0.93042	30
31	.36677	.72649	.39425	.53648	.07491	.93031	29
32	.36704	.72448	.39458	.53432	.07503	.93020	28
33	.36731	.72247	.39492	.53217	.07516	.93010	27
34	.36758	.72047	.39526	.53001	.07528	.92999	26
35	0.36785	2.71847	0.39559	2.52786	1.07540	0.92988	25
36	.36812	.71647	.39593	.52571	.07553	.92978	24
37	.36839	.71448	.39626	.52357	.07565	.92967	23
38	.36867	.71249	.39660	.52142	.07578	.92956	22
39	.36894	.71050	.39694	.51929	.07590	.92945	21
40	0.36921	2.70851	0.39727	2.51715	1.07602	0.92935	20
41	.36948	.70653	.39761	.51502	.07615	.92924	19
42	.36975	.70455	.39795	.51289	.07627	.92913	18
43	.37002	.70258	.39829	.51076	.07640	.92902	17
44	.37029	.70061	.39862	.50864	.07652	.92892	16
45	0.37056	2.69864	0.39896	2.50652	1.07665	0.92881	15
46	.37083	.69667	.39930	.50440	.07677	.92870	14
47	.37110	.69471	.39963	.50229	.07690	.92859	13
48	.37137	.69275	.39997	.50018	.07702	.92849	12
49	.37164	.69079	.40031	.49807	.07715	.92838	11
50	0.37191	2.68884	0.40065	2.49597	1.07727	0.92827	10
51	.37218	.68689	.40098	.49386	.07740	.92816	9
52	.37245	.68494	.40132	.49177	.07752	.92805	8
53	.37272	.68299	.40166	.48967	.07765	.92794	7
54	.37299	.68105	.40200	.48758	.07778	.92784	6
55	0.37326	2.67911	0.40234	2.48549	1.07790	0.92773	5
56	.37353	.67718	.40267	.48340	.07803	.92762	4
57	.37380	.67525	.40301	.48132	.07816	.92751	3
58	.37407	.67332	.40335	.47924	.07828	.92740	2
59	.37434	.67139	.40369	.47716	.07841	.92729	1
60	0.37461	2.66947	0.40403	2.47509	1.07853	0.92718	0
111°→	cos	sec	cot	tan	csc	sin	←68°

22°→	sin	csc	tan	cot	sec	cos	←157°
'							'
0	0.37461	2.66947	0.40403	2.47509	1.07853	0.92718	60
1	.37488	.66755	.40436	.47302	.07866	.92707	59
2	.37515	.66563	.40470	.47095	.07879	.92697	58
3	.37542	.66371	.40504	.46888	.07892	.92686	57
4	.37569	.66180	.40538	.46682	.07904	.92675	56
5	0.37595	2.65989	0.40572	2.46476	1.07917	0.92664	55
6	.37622	.65799	.40606	.46270	.07930	.92653	54
7	.37649	.65609	.40640	.46065	.07943	.92642	53
8	.37676	.65419	.40674	.45860	.07955	.92631	52
9	.37703	.65229	.40707	.45655	.07968	.92620	51
10	0.37730	2.65040	0.40741	2.45451	1.07981	0.92609	50
11	.37757	.64851	.40775	.45246	.07994	.92598	49
12	.37784	.64662	.40809	.45043	.08006	.92587	48
13	.37811	.64473	.40843	.44839	.08019	.92576	47
14	.37838	.64285	.40877	.44636	.08032	.92565	46
15	0.37865	2.64097	0.40911	2.44433	1.08045	0.92554	45
16	.37892	.63909	.40945	.44230	.08058	.92543	44
17	.37919	.63722	.40979	.44027	.08071	.92532	43
18	.37946	.63535	.41013	.43825	.08084	.92521	42
19	.37973	.63348	.41047	.43623	.08097	.92510	41
20	0.37999	2.63162	0.41081	2.43422	1.08109	0.92499	40
21	.38026	.62976	.41115	.43220	.08122	.92488	39
22	.38053	.62790	.41149	.43019	.08135	.92477	38
23	.38080	.62604	.41183	.42819	.08148	.92466	37
24	.38107	.62419	.41217	.42618	.08161	.92455	36
25	0.38134	2.62234	0.41251	2.42418	1.08174	0.92444	35
26	.38161	.62049	.41285	.42218	.08187	.92432	34
27	.38188	.61864	.41319	.42019	.08200	.92421	33
28	.38215	.61680	.41353	.41819	.08213	.92410	32
29	.38241	.61496	.41387	.41620	.08226	.92399	31
30	0.38268	2.61313	0.41421	2.41421	1.08239	0.92388	30
31	.38295	.61129	.41455	.41223	.08252	.92377	29
32	.38322	.60946	.41490	.41025	.08265	.92366	28
33	.38349	.60763	.41524	.40827	.08278	.92355	27
34	.38376	.60581	.41558	.40629	.08291	.92343	26
35	0.38403	2.60399	0.41592	2.40432	1.08305	0.92332	25
36	.38430	.60217	.41626	.40235	.08318	.92321	24
37	.38456	.60035	.41660	.40038	.08331	.92310	23
38	.38483	.59853	.41694	.39841	.08344	.92299	22
39	.38510	.59672	.41728	.39645	.08357	.92287	21
40	0.38537	2.59491	0.41763	2.39449	1.08370	0.92276	20
41	.38564	.59311	.41797	.39253	.08383	.92265	19
42	.38591	.59130	.41831	.39058	.08397	.92254	18
43	.38617	.58950	.41865	.38863	.08410	.92243	17
44	.38644	.58771	.41899	.38668	.08423	.92231	16
45	0.38671	2.58591	0.41933	2.38473	1.08436	0.92220	15
46	.38698	.58412	.41968	.38279	.08449	.92209	14
47	.38725	.58233	.42002	.38084	.08463	.92198	13
48	.38752	.58054	.42036	.37891	.08476	.92186	12
49	.38778	.57876	.42070	.37697	.08489	.92175	11
50	0.38805	2.57698	0.42105	2.37504	1.08503	0.92164	10
51	.38832	.57520	.42139	.37311	.08516	.92152	9
52	.38859	.57342	.42173	.37118	.08529	.92141	8
53	.38886	.57165	.42207	.36925	.08542	.92130	7
54	.38912	.56988	.42242	.36733	.08556	.92119	6
55	0.38939	2.56811	0.42276	2.36541	1.08569	0.92107	5
56	.38966	.56634	.42310	.36349	.08582	.92096	4
57	.38993	.56458	.42345	.36158	.08596	.92085	3
58	.39020	.56282	.42379	.35967	.08609	.92073	2
59	.39046	.56106	.42413	.35776	.08623	.92062	1
60	0.39073	2.55930	0.42447	2.35585	1.08636	0.92050	0
112°→	cos	sec	cot	tan	csc	sin	←67°

23°→	sin	csc	tan	cot	sec	cos	←156°
'							'
0	0.39073	2.55930	0.42447	2.35585	1.08636	0.92050	60
1	.39100	.55755	.42482	.35395	.08649	.92039	59
2	.39127	.55580	.42516	.35205	.08663	.92028	58
3	.39153	.55405	.42551	.35015	.08676	.92016	57
4	.39180	.55231	.42585	.34825	.08690	.92005	56
5	0.39207	2.55057	0.42619	2.34636	1.08703	0.91994	55
6	.39234	.54883	.42654	.34447	.08717	.91982	54
7	.39260	.54709	.42689	.34258	.08730	.91971	53
8	.39287	.54536	.42722	.34069	.08744	.91959	52
9	.39314	.54363	.42757	.33881	.08757	.91948	51
10	0.39341	2.54190	0.42791	2.33693	1.08771	0.91936	50
11	.39367	.54017	.42826	.33505	.08784	.91925	49
12	.39394	.53845	.42860	.33317	.08798	.91914	48
13	.39421	.53672	.42894	.33130	.08811	.91902	47
14	.39448	.53500	.42929	.32943	.08825	.91891	46
15	0.39474	2.53329	0.42963	2.32756	1.08839	0.91879	45
16	.39501	.53157	.42998	.32570	.08852	.91868	44
17	.39528	.52986	.43032	.32383	.08866	.91856	43
18	.39555	.52815	.43067	.32197	.08880	.91845	42
19	.39581	.52645	.43101	.32012	.08893	.91833	41
20	0.39608	2.52474	0.43136	2.31826	1.08907	0.91822	40
21	.39635	.52304	.43170	.31641	.08920	.91810	39
22	.39661	.52134	.43205	.31456	.08934	.91799	38
23	.39688	.51965	.43239	.31271	.08948	.91787	37
24	.39715	.51795	.43274	.31086	.08962	.91775	36
25	0.39741	2.51626	0.43308	2.30902	1.08975	0.91764	35
26	.39768	.51457	.43343	.30718	.08989	.91752	34
27	.39795	.51289	.43378	.30534	.09003	.91741	33
28	.39822	.51120	.43412	.30351	.09017	.91729	32
29	.39848	.50952	.43447	.30167	.09030	.91718	31
30	0.39875	2.50784	0.43481	2.29984	1.09044	0.91706	30
31	.39902	.50617	.43516	.29801	.09058	.91694	29
32	.39928	.50449	.43550	.29619	.09072	.91683	28
33	.39955	.50282	.43585	.29437	.09086	.91671	27
34	.39982	.50115	.43620	.29254	.09099	.91660	26
35	0.40008	2.49948	0.43654	2.29073	1.09113	0.91648	25
36	.40035	.49782	.43689	.28891	.09127	.91636	24
37	.40062	.49616	.43724	.28710	.09141	.91625	23
38	.40088	.49450	.43758	.28528	.09155	.91613	22
39	.40115	.49284	.43793	.28348	.09169	.91601	21
40	0.40141	2.49119	0.43828	2.28167	1.09183	0.91590	20
41	.40168	.48954	.43862	.27987	.09197	.91578	19
42	.40195	.48789	.43897	.27806	.09211	.91566	18
43	.40221	.48624	.43932	.27626	.09224	.91555	17
44	.40248	.48459	.43966	.27447	.09238	.91543	16
45	0.40275	2.48295	0.44001	2.27267	1.09252	0.91531	15
46	.40301	.48131	.44036	.27088	.09266	.91519	14
47	.40328	.47967	.44071	.26909	.09280	.91508	13
48	.40355	.47804	.44105	.26730	.09294	.91496	12
49	.40381	.47640	.44140	.26552	.09308	.91484	11
50	0.40408	2.47477	0.44175	2.26374	1.09323	0.91472	10
51	.40434	.47314	.44210	.26196	.09337	.91461	9
52	.40461	.47152	.44244	.26018	.09351	.91449	8
53	.40488	.46989	.44279	.25840	.09365	.91437	7
54	.40514	.46827	.44314	.25663	.09379	.91425	6
55	0.40541	2.46665	0.44349	2.25486	1.09393	0.91414	5
56	.40567	.46504	.44384	.25309	.09407	.91402	4
57	.40594	.46342	.44418	.25132	.09421	.91390	3
58	.40621	.46181	.44453	.24956	.09435	.91378	2
59	.40647	.46020	.44488	.24780	.09449	.91366	1
60	0.40674	2.45859	0.44523	2.24604	1.09464	0.91355	0
113°→	cos	sec	cot	tan	csc	sin	←66°

24°→ ↓	sin	csc	tan	cot	sec	←155° cos ↓	
0	0.40674	2.45859	0.44523	2.24604	1.09464	0.91355	60
1	.40700	.45699	.44558	.24428	.09478	.91343	59
2	.40727	.45539	.44593	.24252	.09492	.91331	58
3	.40753	.45378	.44627	.24077	.09506	.91319	57
4	.40780	.45219	.44662	.23902	.09520	.91307	56
5	0.40806	2.45059	0.44697	2.23727	1.09535	0.91295	55
6	.40833	.44900	.44732	.23553	.09549	.91283	54
7	.40860	.44741	.44767	.23378	.09563	.91272	53
8	.40886	.44582	.44802	.23204	.09577	.91260	52
9	.40913	.44423	.44837	.23030	.09592	.91248	51
10	0.40939	2.44264	0.44872	2.22857	1.09606	0.91236	50
11	.40966	.44106	.44907	.22683	.09620	.91224	49
12	.40992	.43948	.44942	.22510	.09635	.91212	48
13	.41019	.43790	.44977	.22337	.09649	.91200	47
14	.41045	.43633	.45012	.22164	.09663	.91188	46
15	0.41072	2.43476	0.45047	2.21992	1.09678	0.91176	45
16	.41098	.43318	.45082	.21819	.09692	.91164	44
17	.41125	.43162	.45117	.21647	.09707	.91152	43
18	.41151	.43005	.45152	.21475	.09721	.91140	42
19	.41178	.42848	.45187	.21304	.09735	.91128	41
20	0.41204	2.42692	0.45222	2.21132	1.09750	0.91116	40
21	.41231	.42536	.45257	.20961	.09764	.91104	39
22	.41257	.42380	.45292	.20790	.09779	.91092	38
23	.41284	.42225	.45327	.20619	.09793	.91080	37
24	.41310	.42070	.45362	.20449	.09808	.91068	36
25	0.41337	2.41914	0.45397	2.20278	1.09822	0.91056	35
26	.41363	.41760	.45432	.20108	.09837	.91044	34
27	.41390	.41605	.45467	.19938	.09851	.91032	33
28	.41416	.41450	.45502	.19769	.09866	.91020	32
29	.41443	.41296	.45538	.19599	.09880	.91008	31
30	0.41469	2.41142	0.45573	2.19430	1.09895	0.90996	30
31	.41496	.40988	.45608	.19261	.09909	.90984	29
32	.41522	.40835	.45643	.19092	.09924	.90972	28
33	.41549	.40681	.45678	.18923	.09939	.90960	27
34	.41575	.40528	.45713	.18755	.09953	.90948	26
35	0.41602	2.40375	0.45748	2.18587	1.09968	0.90936	25
36	.41628	.40222	.45784	.18419	.09982	.90924	24
37	.41655	.40070	.45819	.18251	.09997	.90911	23
38	.41681	.39918	.45854	.18084	.10012	.90899	22
39	.41707	.39766	.45889	.17916	.10026	.90887	21
40	0.41734	2.39614	0.45924	2.17749	1.10041	0.90875	20
41	.41760	.39462	.45960	.17582	.10056	.90863	19
42	.41787	.39311	.45995	.17416	.10071	.90851	18
43	.41813	.39159	.46030	.17249	.10085	.90839	17
44	.41840	.39008	.46065	.17083	.10100	.90826	16
45	0.41866	2.38857	0.46101	2.16917	1.10115	0.90814	15
46	.41892	.38707	.46136	.16751	.10130	.90802	14
47	.41919	.38556	.46171	.16585	.10144	.90790	13
48	.41945	.38406	.46206	.16420	.10159	.90778	12
49	.41972	.38256	.46242	.16255	.10174	.90766	11
50	0.41998	2.38106	0.46277	2.16090	1.10189	0.90753	10
51	.42024	.37957	.46312	.15925	.10204	.90741	9
52	.42051	.37808	.46348	.15760	.10218	.90729	8
53	.42077	.37658	.46383	.15596	.10233	.90717	7
54	.42104	.37509	.46418	.15432	.10248	.90704	6
55	0.42130	2.37361	0.46454	2.15268	1.10263	0.90692	5
56	.42156	.37212	.46489	.15104	.10278	.90680	4
57	.42183	.37064	.46525	.14940	.10293	.90668	3
58	.42209	.36916	.46560	.14777	.10308	.90655	2
59	.42235	.36768	.46595	.14614	.10323	.90643	1
60	0.42262	2.36620	0.46631	2.14451	1.10338	0.90631	0
↑ 114°→	cos	sec	cot	tan	csc	sin ←65° ↑	

25°→ ↓	sin	csc	tan	cot	sec	←154° cos ↓	
0	0.42262	2.36620	0.46631	2.14451	1.10338	0.90631	60
1	.42288	.36473	.46666	.14288	.10353	.90618	59
2	.42315	.36325	.46702	.14125	.10368	.90606	58
3	.42341	.36178	.46737	.13963	.10383	.90594	57
4	.42367	.36031	.46772	.13801	.10398	.90582	56
5	0.42394	2.35885	0.46808	2.13639	1.10413	0.90569	55
6	.42420	.35738	.46843	.13477	.10428	.90557	54
7	.42446	.35592	.46879	.13316	.10443	.90545	53
8	.42473	.35446	.46914	.13154	.10458	.90532	52
9	.42499	.35300	.46950	.12993	.10473	.90520	51
10	0.42525	2.35154	0.46985	2.12832	1.10488	0.90507	50
11	.42552	.35009	.47021	.12671	.10503	.90495	49
12	.42578	.34863	.47056	.12511	.10518	.90483	48
13	.42604	.34718	.47092	.12350	.10533	.90470	47
14	.42631	.34573	.47128	.12190	.10549	.90458	46
15	0.42657	2.34429	0.47163	2.12030	1.10564	0.90446	45
16	.42683	.34284	.47199	.11871	.10579	.90433	44
17	.42709	.34140	.47234	.11711	.10594	.90421	43
18	.42736	.33996	.47270	.11552	.10609	.90408	42
19	.42762	.33852	.47305	.11392	.10625	.90396	41
20	0.42788	2.33708	0.47341	2.11233	1.10640	0.90383	40
21	.42815	.33565	.47377	.11075	.10655	.90371	39
22	.42841	.33422	.47412	.10916	.10670	.90358	38
23	.42867	.33278	.47448	.10758	.10686	.90346	37
24	.42894	.33135	.47483	.10600	.10701	.90334	36
25	0.42920	2.32993	0.47519	2.10442	1.10716	0.90321	35
26	.42946	.32850	.47555	.10284	.10731	.90309	34
27	.42972	.32708	.47590	.10126	.10747	.90296	33
28	.42999	.32566	.47626	.09969	.10762	.90284	32
29	.43025	.32424	.47662	.09811	.10777	.90271	31
30	0.43051	2.32282	0.47698	2.09654	1.10793	0.90259	30
31	.43077	.32140	.47733	.09498	.10808	.90246	29
32	.43104	.31999	.47769	.09341	.10824	.90233	28
33	.43130	.31858	.47805	.09184	.10839	.90221	27
34	.43156	.31717	.47840	.09028	.10854	.90208	26
35	0.43182	2.31576	0.47876	2.08872	1.10870	0.90196	25
36	.43209	.31436	.47912	.08716	.10885	.90183	24
37	.43235	.31295	.47948	.08560	.10901	.90171	23
38	.43261	.31155	.47984	.08405	.10916	.90158	22
39	.43287	.31015	.48019	.08250	.10932	.90146	21
40	0.43313	2.30875	0.48055	2.08094	1.10947	0.90133	20
41	.43340	.30735	.48091	.07939	.10963	.90120	19
42	.43366	.30596	.48127	.07785	.10978	.90108	18
43	.43392	.30457	.48163	.07630	.10994	.90095	17
44	.43418	.30318	.48198	.07476	.11009	.90082	16
45	0.43445	2.30179	0.48234	2.07321	1.11025	0.90070	15
46	.43471	.30040	.48270	.07167	.11041	.90057	14
47	.43497	.29901	.48306	.07014	.11056	.90045	13
48	.43523	.29763	.48342	.06860	.11072	.90032	12
49	.43549	.29625	.48378	.06706	.11088	.90019	11
50	0.43575	2.29487	0.48414	2.06553	1.11103	0.90007	10
51	.43602	.29349	.48450	.06400	.11119	.89994	9
52	.43628	.29211	.48486	.06247	.11134	.89981	8
53	.43654	.29074	.48521	.06094	.11150	.89968	7
54	.43680	.28937	.48557	.05942	.11166	.89956	6
55	0.43706	2.28800	0.48593	2.05790	1.11181	0.89943	5
56	.43733	.28663	.48629	.05637	.11197	.89930	4
57	.43759	.28526	.48665	.05485	.11213	.89918	3
58	.43785	.28390	.48701	.05333	.11229	.89905	2
59	.43811	.28253	.48737	.05182	.11244	.89892	1
60	0.43837	2.28117	0.48773	2.05030	1.11260	0.89879	0
↑ 115°→	cos	sec	cot	tan	csc	sin ←64° ↑	

26°→	sin	csc	tan	cot	sec	←153° cos	
0	0.43837	2.28117	0.48773	2.05030	1.11260	0.89879	60
1	.43863	.27981	.48809	.04879	.11276	.89867	59
2	.43889	.27845	.48845	.04728	.11292	.89854	58
3	.43916	.27710	.48881	.04577	.11308	.89841	57
4	.43942	.27574	.48917	.04426	.11323	.89828	56
5	0.43968	2.27439	0.48953	2.04276	1.11339	0.89816	55
6	.43994	.27304	.48989	.04125	.11355	.89803	54
7	.44020	.27169	.49026	.03975	.11371	.89790	53
8	.44046	.27035	.49062	.03825	.11387	.89777	52
9	.44072	.26900	.49098	.03675	.11403	.89764	51
10	.44098	2.26766	0.49134	2.03526	1.11419	0.89752	50
11	.44124	.26632	.49170	.03376	.11435	.89739	49
12	.44151	.26498	.49206	.03227	.11451	.89726	48
13	.44177	.26364	.49242	.03078	.11467	.89713	47
14	.44203	.26230	.49278	.02929	.11483	.89700	46
15	0.44229	2.26097	0.49315	2.02780	1.11499	0.89687	45
16	.44255	.25963	.49351	.02631	.11515	.89674	44
17	.44281	.25830	.49387	.02483	.11531	.89662	43
18	.44307	.25697	.49423	.02335	.11547	.89649	42
19	.44333	.25565	.49459	.02187	.11563	.89636	41
20	0.44359	2.25432	0.49495	2.02039	1.11579	0.89623	40
21	.44385	.25300	.49532	.01891	.11595	.89610	39
22	.44411	.25167	.49568	.01743	.11611	.89597	38
23	.44437	.25035	.49604	.01596	.11627	.89584	37
24	.44464	.24903	.49640	.01449	.11643	.89571	36
25	0.44490	2.24772	0.49677	2.01302	1.11659	0.89558	35
26	.44516	.24640	.49713	.01155	.11675	.89545	34
27	.44542	.24509	.49749	.01008	.11691	.89532	33
28	.44568	.24378	.49786	.00862	.11708	.89519	32
29	.44594	.24247	.49822	.00715	.11724	.89506	31
30	0.44620	2.24116	0.49858	2.00569	1.11740	0.89493	30
31	.44646	.23985	.49894	.00423	.11756	.89480	29
32	.44672	.23855	.49931	.00277	.11772	.89467	28
33	.44698	.23724	.49967	2.00131	.11789	.89454	27
34	.44724	.23594	.50004	1.99986	.11805	.89441	26
35	0.44750	2.23464	0.50040	1.99841	1.11821	0.89428	25
36	.44776	.23334	.50076	.99695	.11838	.89415	24
37	.44802	.23205	.50113	.99550	.11854	.89402	23
38	.44828	.23075	.50149	.99406	.11870	.89389	22
39	.44854	.22946	.50185	.99261	.11886	.89376	21
40	0.44880	2.22817	0.50222	1.99116	1.11903	0.89363	20
41	.44906	.22688	.50258	.98972	.11919	.89350	19
42	.44932	.22559	.50295	.98828	.11936	.89337	18
43	.44958	.22430	.50331	.98684	.11952	.89324	17
44	.44984	.22302	.50368	.98540	.11968	.89311	16
45	0.45010	2.22174	0.50404	1.98396	1.11985	0.89298	15
46	.45036	.22045	.50441	.98253	.12001	.89285	14
47	.45062	.21918	.50477	.98110	.12018	.89272	13
48	.45088	.21790	.50514	.97966	.12034	.89259	12
49	.45114	.21662	.50550	.97823	.12051	.89245	11
50	0.45140	2.21535	0.50587	1.97681	1.12067	0.89232	10
51	.45166	.21407	.50623	.97538	.12083	.89219	9
52	.45192	.21280	.50660	.97395	.12100	.89206	8
53	.45218	.21153	.50696	.97253	.12117	.89193	7
54	.45243	.21026	.50733	.97111	.12133	.89180	6
55	0.45269	2.20900	0.50769	1.96969	1.12150	0.89167	5
56	.45295	.20773	.50806	.96827	.12166	.89153	4
57	.45321	.20647	.50843	.96685	.12183	.89140	3
58	.45347	.20521	.50879	.96544	.12199	.89127	2
59	.45373	.20395	.50916	.96402	.12216	.89114	1
60	0.45399	2.20269	0.50953	1.96261	1.12233	0.89101	0
116°→ cos	sec	cot	tan	csc	sin ←63°		

27°→	sin	csc	tan	cot	sec	←152° cos	
0	0.45399	2.20269	0.50953	1.96261	1.12233	0.89101	60
1	.45425	.20143	.50989	.96120	.12249	.89087	59
2	.45451	.20018	.51026	.95979	.12266	.89074	58
3	.45477	.19892	.51063	.95838	.12283	.89061	57
4	.45503	.19767	.51099	.95698	.12299	.89048	56
5	0.45529	2.19642	0.51136	1.95557	1.12316	0.89035	55
6	.45554	.19517	.51173	.95417	.12333	.89021	54
7	.45580	.19393	.51209	.95277	.12349	.89008	53
8	.45606	.19268	.51246	.95137	.12366	.88995	52
9	.45632	.19144	.51283	.94997	.12383	.88981	51
10	0.45658	2.19019	0.51319	1.94858	1.12400	0.88968	50
11	.45684	.18895	.51356	.94718	.12416	.88955	49
12	.45710	.18772	.51393	.94579	.12433	.88942	48
13	.45736	.18648	.51430	.94440	.12450	.88928	47
14	.45762	.18524	.51467	.94301	.12467	.88915	46
15	0.45787	2.18401	0.51503	1.94162	1.12484	0.88902	45
16	.45813	.18277	.51540	.94023	.12501	.88888	44
17	.45839	.18154	.51577	.93885	.12518	.88875	43
18	.45865	.18031	.51614	.93746	.12534	.88862	42
19	.45891	.17909	.51651	.93608	.12551	.88848	41
20	0.45917	2.17786	0.51688	1.93470	1.12568	0.88835	40
21	.45942	.17663	.51724	.93332	.12585	.88822	39
22	.45968	.17541	.51761	.93195	.12602	.88808	38
23	.45994	.17419	.51798	.93057	.12619	.88795	37
24	.46020	.17297	.51835	.92920	.12636	.88782	36
25	0.46046	2.17175	0.51872	1.92782	1.12653	0.88768	35
26	.46072	.17053	.51909	.92645	.12670	.88755	34
27	.46097	.16932	.51946	.92508	.12687	.88741	33
28	.46123	.16810	.51983	.92371	.12704	.88728	32
29	.46149	.16689	.52020	.92235	.12721	.88715	31
30	0.46175	2.16568	0.52057	1.92098	1.12738	0.88701	30
31	.46201	.16447	.52094	.91962	.12755	.88688	29
32	.46226	.16326	.52131	.91826	.12772	.88674	28
33	.46252	.16206	.52168	.91690	.12789	.88661	27
34	.46278	.16085	.52205	.91554	.12807	.88647	26
35	0.46304	2.15965	0.52242	1.91418	1.12824	0.88634	25
36	.46330	.15845	.52279	.91282	.12841	.88620	24
37	.46355	.15725	.52316	.91147	.12858	.88607	23
38	.46381	.15605	.52353	.91012	.12875	.88593	22
39	.46407	.15485	.52390	.90876	.12892	.88580	21
40	0.46433	2.15366	0.52427	1.90741	1.12910	0.88566	20
41	.46458	.15246	.52464	.90607	.12927	.88553	19
42	.46484	.15127	.52501	.90472	.12944	.88539	18
43	.46510	.15008	.52538	.90337	.12961	.88526	17
44	.46536	.14889	.52575	.90203	.12979	.88512	16
45	0.46561	2.14770	0.52613	1.90069	1.12996	0.88499	15
46	.46587	.14651	.52650	.89935	.13013	.88485	14
47	.46613	.14533	.52687	.89801	.13031	.88472	13
48	.46639	.14414	.52724	.89667	.13048	.88458	12
49	.46664	.14296	.52761	.89533	.13065	.88445	11
50	0.46690	2.14178	0.52798	1.89400	1.13083	0.88431	10
51	.46716	.14060	.52836	.89266	.13100	.88417	9
52	.46742	.13942	.52873	.89133	.13117	.88404	8
53	.46767	.13825	.52910	.89000	.13135	.88390	7
54	.46793	.13707	.52947	.88867	.13152	.88377	6
55	0.46819	2.13590	0.52985	1.88734	1.13170	0.88363	5
56	.46844	.13473	.53022	.88602	.13187	.88349	4
57	.46870	.13356	.53059	.88469	.13205	.88336	3
58	.46896	.13239	.53096	.88337	.13222	.88322	2
59	.46921	.13122	.53134	.88205	.13239	.88308	1
60	0.46947	2.13005	0.53171	1.88073	1.13257	0.88295	0
117°→ cos	sec	cot	tan	csc	sin ←62°		

28° → ←151°

'	sin	csc	tan	cot	sec	cos	
0	0.46947	2.13005	0.53171	1.88073	1.13257	0.88295	60
1	.46973	.12889	.53208	.87941	.13275	.88281	59
2	.46999	.12773	.53246	.87809	.13292	.88267	58
3	.47024	.12657	.53283	.87677	.13310	.88254	57
4	.47050	.12540	.53320	.87546	.13327	.88240	56
5	0.47076	2.12425	0.53358	1.87415	1.13345	0.88226	55
6	.47101	.12309	.53395	.87283	.13362	.88213	54
7	.47127	.12193	.53432	.87152	.13380	.88199	53
8	.47153	.12078	.53470	.87021	.13398	.88185	52
9	.47178	.11963	.53507	.86891	.13415	.88172	51
10	0.47204	2.11847	0.53545	1.86760	1.13433	0.88158	50
11	.47229	.11732	.53582	.86630	.13451	.88144	49
12	.47255	.11617	.53620	.86499	.13468	.88130	48
13	.47281	.11503	.53657	.86369	.13486	.88117	47
14	.47306	.11388	.53694	.86239	.13504	.88103	46
15	0.47332	2.11274	0.53732	1.86109	1.13521	0.88089	45
16	.47358	.11159	.53769	.85979	.13539	.88075	44
17	.47383	.11045	.53807	.85850	.13557	.88062	43
18	.47409	.10931	.53844	.85720	.13575	.88048	42
19	.47434	.10817	.53882	.85591	.13593	.88034	41
20	0.47460	2.10704	0.53920	1.85462	1.13610	0.88020	40
21	.47486	.10590	.53957	.85333	.13628	.88006	39
22	.47511	.10477	.53995	.85204	.13646	.87993	38
23	.47537	.10363	.54032	.85075	.13664	.87979	37
24	.47562	.10250	.54070	.84946	.13682	.87965	36
25	0.47588	2.10137	0.54107	1.84818	1.13700	0.87951	35
26	.47614	.10024	.54145	.84689	.13718	.87937	34
27	.47639	.09911	.54183	.84561	.13735	.87923	33
28	.47665	.09799	.54220	.84433	.13753	.87909	32
29	.47690	.09686	.54258	.84305	.13771	.87896	31
30	0.47716	2.09574	0.54296	1.84177	1.13789	0.87882	30
31	.47741	.09462	.54333	.84049	.13807	.87868	29
32	.47767	.09350	.54371	.83922	.13825	.87854	28
33	.47793	.09238	.54409	.83794	.13843	.87840	27
34	.47818	.09126	.54446	.83667	.13861	.87826	26
35	0.47844	2.09014	0.54484	1.83540	1.13879	0.87812	25
36	.47869	.08903	.54522	.83413	.13897	.87798	24
37	.47895	.08791	.54560	.83286	.13916	.87784	23
38	.47920	.08680	.54597	.83159	.13934	.87770	22
39	.47946	.08569	.54635	.83033	.13952	.87756	21
40	0.47971	2.08458	0.54673	1.82906	1.13970	0.87743	20
41	.47997	.08347	.54711	.82780	.13988	.87729	19
42	.48022	.08236	.54748	.82654	.14006	.87715	18
43	.48048	.08126	.54786	.82528	.14024	.87701	17
44	.48073	.08015	.54824	.82402	.14042	.87687	16
45	0.48099	2.07905	0.54862	1.82276	1.14061	0.87673	15
46	.48124	.07795	.54900	.82150	.14079	.87659	14
47	.48150	.07685	.54938	.82025	.14097	.87645	13
48	.48175	.07575	.54975	.81899	.14115	.87631	12
49	.48201	.07465	.55013	.81774	.14134	.87617	11
50	0.48226	2.07356	0.55051	1.81649	1.14152	0.87603	10
51	.48252	.07246	.55089	.81524	.14170	.87589	9
52	.48277	.07137	.55127	.81399	.14188	.87575	8
53	.48303	.07027	.55165	.81274	.14207	.87561	7
54	.48328	.06918	.55203	.81150	.14225	.87546	6
55	0.48354	2.06809	0.55241	1.81025	1.14243	0.87532	5
56	.48379	.06701	.55279	.80901	.14262	.87518	4
57	.48405	.06592	.55317	.80777	.14280	.87504	3
58	.48430	.06483	.55355	.80653	.14299	.87490	2
59	.48456	.06375	.55393	.80529	.14317	.87476	1
60	0.48481	2.06267	0.55431	1.80405	1.14335	0.87462	0

118°→ cos | sec | cot | tan | csc | sin ←61°

29° → ←150°

'	sin	csc	tan	cot	sec	cos	
0	0.48481	2.06267	0.55431	1.80405	1.14335	0.87462	60
1	.48506	.06158	.55469	.80281	.14354	.87448	59
2	.48532	.06050	.55507	.80158	.14372	.87434	58
3	.48557	.05942	.55545	.80034	.14391	.87420	57
4	.48583	.05835	.55583	.79911	.14409	.87406	56
5	0.48608	2.05727	0.55621	1.79788	1.14428	0.87391	55
6	.48634	.05619	.55659	.79665	.14446	.87377	54
7	.48659	.05512	.55697	.79542	.14465	.87363	53
8	.48684	.05405	.55736	.79419	.14483	.87349	52
9	.48710	.05298	.55774	.79296	.14502	.87335	51
10	0.48735	2.05191	0.55812	1.79174	1.14521	0.87321	50
11	.48761	.05084	.55850	.79051	.14539	.87306	49
12	.48786	.04977	.55888	.78929	.14558	.87292	48
13	.48811	.04870	.55926	.78807	.14576	.87278	47
14	.48837	.04764	.55964	.78685	.14595	.87264	46
15	0.48862	2.04657	0.56003	1.78563	1.14614	0.87250	45
16	.48888	.04551	.56041	.78441	.14632	.87235	44
17	.48913	.04445	.56079	.78319	.14651	.87221	43
18	.48938	.04339	.56117	.78198	.14670	.87207	42
19	.48964	.04233	.56156	.78077	.14689	.87193	41
20	0.48989	2.04128	0.56194	1.77955	1.14707	0.87178	40
21	.49014	.04022	.56232	.77834	.14726	.87164	39
22	.49040	.03916	.56270	.77713	.14745	.87150	38
23	.49065	.03811	.56309	.77592	.14764	.87136	37
24	.49090	.03706	.56347	.77471	.14782	.87121	36
25	0.49116	2.03601	0.56385	1.77351	1.14801	0.87107	35
26	.49141	.03496	.56424	.77230	.14820	.87093	34
27	.49166	.03391	.56462	.77110	.14839	.87079	33
28	.49192	.03286	.56501	.76990	.14858	.87064	32
29	.49217	.03182	.56539	.76869	.14877	.87050	31
30	0.49242	2.03077	0.56577	1.76749	1.14896	0.87036	30
31	.49268	.02973	.56616	.76629	.14914	.87021	29
32	.49293	.02869	.56654	.76510	.14933	.87007	28
33	.49318	.02765	.56693	.76390	.14952	.86993	27
34	.49344	.02661	.56731	.76271	.14971	.86978	26
35	0.49369	2.02557	0.56769	1.76151	1.14990	0.86964	25
36	.49394	.02453	.56808	.76032	.15009	.86949	24
37	.49419	.02349	.56846	.75913	.15028	.86935	23
38	.49445	.02246	.56885	.75794	.15047	.86921	22
39	.49470	.02143	.56923	.75675	.15066	.86906	21
40	0.49495	2.02039	0.56962	1.75556	1.15085	0.86892	20
41	.49521	.01936	.57000	.75437	.15105	.86878	19
42	.49546	.01833	.57039	.75319	.15124	.86863	18
43	.49571	.01730	.57078	.75200	.15143	.86849	17
44	.49596	.01628	.57116	.75082	.15162	.86834	16
45	0.49622	2.01525	0.57155	1.74964	1.15181	0.86820	15
46	.49647	.01422	.57193	.74846	.15200	.86805	14
47	.49672	.01320	.57232	.74728	.15219	.86791	13
48	.49697	.01218	.57271	.74610	.15239	.86777	12
49	.49723	.01116	.57309	.74492	.15258	.86762	11
50	0.49748	2.01014	0.57348	1.74375	1.15277	0.86748	10
51	.49773	.00912	.57386	.74257	.15296	.86733	9
52	.49798	.00810	.57425	.74140	.15315	.86719	8
53	.49824	.00708	.57464	.74022	.15335	.86704	7
54	.49849	.00607	.57503	.73905	.15354	.86690	6
55	0.49874	2.00505	0.57541	1.73788	1.15373	0.86675	5
56	.49899	.00404	.57580	.73671	.15393	.86661	4
57	.49924	.00303	.57619	.73555	.15412	.86646	3
58	.49950	.00202	.57657	.73438	.15431	.86632	2
59	.49975	.00101	.57696	.73321	.15451	.86617	1
60	0.50000	2.00000	0.57735	1.73205	1.15470	0.86603	0

119°→ cos | sec | cot | tan | csc | sin ←60°

30°→	sin	csc	tan	cot	sec	←149° cos	
′							′
0	0.50000	2.00000	0.57735	1.73205	1.15470	0.86603	60
1	.50025	1.99899	.57774	.73089	.15489	.86588	59
2	.50050	.99799	.57813	.72973	.15509	.86573	58
3	.50076	.99698	.57851	.72857	.15528	.86559	57
4	.50101	.99598	.57890	.72741	.15548	.86544	56
5	0.50126	1.99498	0.57929	1.72625	1.15567	0.86530	55
6	.50151	.99398	.57968	.72509	.15587	.86515	54
7	.50176	.99298	.58007	.72393	.15606	.86501	53
8	.50201	.99198	.58046	.72278	.15626	.86486	52
9	.50227	.99098	.58085	.72163	.15645	.86471	51
10	0.50252	1.98998	0.58124	1.72047	1.15665	0.86457	50
11	.50277	.98899	.58162	.71932	.15684	.86442	49
12	.50302	.98799	.58201	.71817	.15704	.86427	48
13	.50327	.98700	.58240	.71702	.15724	.86413	47
14	.50352	.98601	.58279	.71588	.15743	.86398	46
15	0.50377	1.98502	0.58318	1.71473	1.15763	0.86384	45
16	.50403	.98403	.58357	.71358	.15782	.86369	44
17	.50428	.98304	.58396	.71244	.15802	.86354	43
18	.50453	.98205	.58435	.71129	.15822	.86340	42
19	.50478	.98107	.58474	.71015	.15841	.86325	41
20	0.50503	1.98008	0.58513	1.70901	1.15861	0.86310	40
21	.50528	.97910	.58552	.70787	.15881	.86295	39
22	.50553	.97811	.58591	.70673	.15901	.86281	38
23	.50578	.97713	.58631	.70560	.15920	.86266	37
24	.50603	.97615	.58670	.70446	.15940	.86251	36
25	0.50628	1.97517	0.58709	1.70332	1.15960	0.86237	35
26	.50654	.97420	.58748	.70219	.15980	.86222	34
27	.50679	.97322	.58787	.70106	.16000	.86207	33
28	.50704	.97224	.58826	.69992	.16019	.86192	32
29	.50729	.97127	.58865	.69879	.16039	.86178	31
30	0.50754	1.97029	0.58905	1.69766	1.16059	0.86163	30
31	.50779	.96932	.58944	.69653	.16079	.86148	29
32	.50804	.96835	.58983	.69541	.16099	.86133	28
33	.50829	.96738	.59022	.69428	.16119	.86119	27
34	.50854	.96641	.59061	.69316	.16139	.86104	26
35	0.50879	1.96544	0.59101	1.69203	1.16159	0.86089	25
36	.50904	.96448	.59140	.69091	.16179	.86074	24
37	.50929	.96351	.59179	.68979	.16199	.86059	23
38	.50954	.96255	.59218	.68866	.16219	.86045	22
39	.50979	.96158	.59258	.68754	.16239	.86030	21
40	0.51004	1.96062	0.59297	1.68643	1.16259	0.86015	20
41	.51029	.95966	.59336	.68531	.16279	.86000	19
42	.51054	.95870	.59376	.68419	.16299	.85985	18
43	.51079	.95774	.59415	.68308	.16319	.85970	17
44	.51104	.95678	.59454	.68196	.16339	.85956	16
45	0.51129	1.95583	0.59494	1.68085	1.16359	0.85941	15
46	.51154	.95487	.59533	.67974	.16380	.85926	14
47	.51179	.95392	.59573	.67863	.16400	.85911	13
48	.51204	.95296	.59612	.67752	.16420	.85896	12
49	.51229	.95201	.59651	.67641	.16440	.85881	11
50	0.51254	1.95106	0.59691	1.67530	1.16460	0.85866	10
51	.51279	.95011	.59730	.67419	.16481	.85851	9
52	.51304	.94916	.59770	.67309	.16501	.85836	8
53	.51329	.94821	.59809	.67198	.16521	.85821	7
54	.51354	.94726	.59849	.67088	.16541	.85806	6
55	0.51379	1.94632	0.59888	1.66978	1.16562	0.85792	5
56	.51404	.94537	.59928	.66867	.16582	.85777	4
57	.51429	.94443	.59967	.66757	.16602	.85762	3
58	.51454	.94349	.60007	.66647	.16623	.85747	2
59	.51479	.94254	.60046	.66538	.16643	.85732	1
60	0.51504	1.94160	0.60086	1.66428	1.16663	0.85717	0
↑120°→	cos	sec	cot	tan	csc	sin ←59°↑	

31°→	sin	csc	tan	cot	sec	←148° cos	
′							′
0	0.51504	1.94160	0.60086	1.66428	1.16663	0.85717	60
1	.51529	.94066	.60126	.66318	.16684	.85702	59
2	.51554	.93973	.60165	.66209	.16704	.85687	58
3	.51579	.93879	.60205	.66099	.16725	.85672	57
4	.51604	.93785	.60245	.65990	.16745	.85657	56
5	0.51628	1.93692	0.60284	1.65881	1.16766	0.85642	55
6	.51653	.93598	.60324	.65772	.16786	.85627	54
7	.51678	.93505	.60364	.65663	.16806	.85612	53
8	.51703	.93412	.60403	.65554	.16827	.85597	52
9	.51728	.93319	.60443	.65445	.16848	.85582	51
10	0.51753	1.93226	0.60483	1.65337	1.16868	0.85567	50
11	.51778	.93133	.60522	.65228	.16889	.85551	49
12	.51803	.93040	.60562	.65120	.16909	.85536	48
13	.51828	.92947	.60602	.65011	.16930	.85521	47
14	.51852	.92855	.60642	.64903	.16950	.85506	46
15	0.51877	1.92762	0.60681	1.64795	1.16971	0.85491	45
16	.51902	.92670	.60721	.64687	.16992	.85476	44
17	.51927	.92578	.60761	.64579	.17012	.85461	43
18	.51952	.92486	.60801	.64471	.17033	.85446	42
19	.51977	.92394	.60841	.64363	.17054	.85431	41
20	0.52002	1.92302	0.60881	1.64256	1.17075	0.85416	40
21	.52026	.92210	.60921	.64148	.17095	.85401	39
22	.52051	.92118	.60960	.64041	.17116	.85385	38
23	.52076	.92027	.61000	.63934	.17137	.85370	37
24	.52101	.91935	.61040	.63826	.17158	.85355	36
25	0.52126	1.91844	0.61080	1.63719	1.17178	0.85340	35
26	.52151	.91752	.61120	.63612	.17199	.85325	34
27	.52175	.91661	.61160	.63505	.17220	.85310	33
28	.52200	.91570	.61200	.63398	.17241	.85294	32
29	.52225	.91479	.61240	.63292	.17262	.85279	31
30	0.52250	1.91388	0.61280	1.63185	1.17283	0.85264	30
31	.52275	.91297	.61320	.63079	.17304	.85249	29
32	.52299	.91207	.61360	.62972	.17325	.85234	28
33	.52324	.91116	.61400	.62866	.17346	.85218	27
34	.52349	.91026	.61440	.62760	.17367	.85203	26
35	0.52374	1.90935	0.61480	1.62654	1.17388	0.85188	25
36	.52399	.90845	.61520	.62548	.17409	.85173	24
37	.52423	.90755	.61561	.62442	.17430	.85157	23
38	.52448	.90665	.61601	.62336	.17451	.85142	22
39	.52473	.90575	.61641	.62230	.17472	.85127	21
40	0.52498	1.90485	0.61681	1.62125	1.17493	0.85112	20
41	.52522	.90395	.61721	.62019	.17514	.85096	19
42	.52547	.90305	.61761	.61914	.17535	.85081	18
43	.52572	.90216	.61801	.61809	.17556	.85066	17
44	.52597	.90126	.61842	.61703	.17577	.85051	16
45	0.52621	1.90037	0.61882	1.61598	1.17598	0.85035	15
46	.52646	.89948	.61922	.61493	.17620	.85020	14
47	.52671	.89858	.61962	.61388	.17641	.85005	13
48	.52696	.89769	.62003	.61283	.17662	.84989	12
49	.52720	.89680	.62043	.61179	.17683	.84974	11
50	0.52745	1.89591	0.62083	1.61074	1.17704	0.84959	10
51	.52770	.89503	.62124	.60970	.17726	.84943	9
52	.52794	.89414	.62164	.60865	.17747	.84928	8
53	.52819	.89325	.62204	.60761	.17768	.84913	7
54	.52844	.89237	.62245	.60657	.17790	.84897	6
55	0.52869	1.89148	0.62285	1.60553	1.17811	0.84882	5
56	.52893	.89060	.62325	.60449	.17832	.84866	4
57	.52918	.88972	.62366	.60345	.17854	.84851	3
58	.52943	.88884	.62406	.60241	.17875	.84836	2
59	.52967	.88796	.62446	.60137	.17896	.84820	1
60	0.52992	1.88708	0.62487	1.60033	1.17918	0.84805	0
↑121°→	cos	sec	cot	tan	csc	sin ←58°↑	

32°→	sin	csc	tan	cot	sec	cos ←147°	
′							′
0	0.52992	1.88708	0.62487	1.60033	1.17918	0.84805	60
1	.53017	.88620	.62527	.59930	.17939	.84789	59
2	.53041	.88532	.62568	.59826	.17961	.84774	58
3	.53066	.88445	.62608	.59723	.17982	.84759	57
4	.53091	.88357	.62649	.59620	.18004	.84743	56
5	0.53115	1.88270	0.62689	1.59517	1.18025	0.84728	55
6	.53140	.88183	.62730	.59414	.18047	.84712	54
7	.53164	.88095	.62770	.59311	.18068	.84697	53
8	.53189	.88008	.62811	.59208	.18090	.84681	52
9	.53214	.87921	.62852	.59105	.18111	.84666	51
10	0.53238	1.87834	0.62892	1.59002	1.18133	0.84650	50
11	.53263	.87748	.62933	.58900	.18155	.84635	49
12	.53288	.87661	.62973	.58797	.18176	.84619	48
13	.53312	.87574	.63014	.58695	.18198	.84604	47
14	.53337	.87488	.63055	.58593	.18220	.84588	46
15	0.53361	1.87401	0.63095	1.58490	1.18241	0.84573	45
16	.53386	.87315	.63136	.58388	.18263	.84557	44
17	.53411	.87229	.63177	.58286	.18285	.84542	43
18	.53435	.87142	.63217	.58184	.18307	.84526	42
19	.53460	.87056	.63258	.58083	.18328	.84511	41
20	0.53484	1.86970	0.63299	1.57981	1.18350	0.84495	40
21	.53509	.86885	.63340	.57879	.18372	.84480	39
22	.53534	.86799	.63380	.57778	.18394	.84464	38
23	.53558	.86713	.63421	.57676	.18416	.84448	37
24	.53583	.86627	.63462	.57575	.18437	.84433	36
25	0.53607	1.86542	0.63503	1.57474	1.18459	0.84417	35
26	.53632	.86457	.63544	.57372	.18481	.84402	34
27	.53656	.86371	.63584	.57271	.18503	.84386	33
28	.53681	.86286	.63625	.57170	.18525	.84370	32
29	.53705	.86201	.63666	.57069	.18547	.84355	31
30	0.53730	1.86116	0.63707	1.56969	1.18569	0.84339	30
31	.53754	.86031	.63748	.56868	.18591	.84324	29
32	.53779	.85946	.63789	.56767	.18613	.84308	28
33	.53804	.85861	.63830	.56667	.18635	.84292	27
34	.53828	.85777	.63871	.56566	.18657	.84277	26
35	0.53853	1.85692	0.63912	1.56466	1.18679	0.84261	25
36	.53877	.85608	.63953	.56366	.18701	.84245	24
37	.53902	.85523	.63994	.56265	.18723	.84230	23
38	.53926	.85439	.64035	.56165	.18745	.84214	22
39	.53951	.85355	.64076	.56065	.18767	.84198	21
40	0.53975	1.85271	0.64117	1.55966	1.18790	0.84182	20
41	.54000	.85187	.64158	.55866	.18812	.84167	19
42	.54024	.85103	.64199	.55766	.18834	.84151	18
43	.54049	.85019	.64240	.55666	.18856	.84135	17
44	.54073	.84935	.64281	.55567	.18878	.84120	16
45	0.54097	1.84852	0.64322	1.55467	1.18901	0.84104	15
46	.54122	.84768	.64363	.55368	.18923	.84088	14
47	.54146	.84685	.64404	.55269	.18945	.84072	13
48	.54171	.84601	.64446	.55170	.18967	.84057	12
49	.54195	.84518	.64487	.55071	.18990	.84041	11
50	0.54220	1.84435	0.64528	1.54972	1.19012	0.84025	10
51	.54244	.84352	.64569	.54873	.19034	.84009	9
52	.54269	.84269	.64610	.54774	.19057	.83994	8
53	.54293	.84186	.64652	.54675	.19079	.83978	7
54	.54317	.84103	.64693	.54576	.19102	.83962	6
55	0.54342	1.84020	0.64734	1.54478	1.19124	0.83946	5
56	.54366	.83938	.64775	.54379	.19146	.83930	4
57	.54391	.83855	.64817	.54281	.19169	.83915	3
58	.54415	.83773	.64858	.54183	.19191	.83899	2
59	.54440	.83690	.64899	.54085	.19214	.83883	1
60	0.54464	1.83608	0.64941	1.53986	1.19236	0.83867	0
122°→ cos	sec	cot	tan	csc	sin ←57°		

33°→	sin	csc	tan	cot	sec	cos ←146°	
′							′
0	0.54464	1.83608	0.64941	1.53986	1.19236	0.83867	60
1	.54488	.83526	.64982	.53888	.19259	.83851	59
2	.54513	.83444	.65024	.53791	.19281	.83835	58
3	.54537	.83362	.65065	.53693	.19304	.83819	57
4	.54561	.83280	.65106	.53595	.19327	.83804	56
5	0.54586	1.83198	0.65148	1.53497	1.19349	0.83788	55
6	.54610	.83116	.65189	.53400	.19372	.83772	54
7	.54635	.83034	.65231	.53302	.19394	.83756	53
8	.54659	.82953	.65272	.53205	.19417	.83740	52
9	.54683	.82871	.65314	.53107	.19440	.83724	51
10	0.54708	1.82790	0.65355	1.53010	1.19463	0.83708	50
11	.54732	.82709	.65397	.52913	.19485	.83692	49
12	.54756	.82627	.65438	.52816	.19508	.83676	48
13	.54781	.82546	.65480	.52719	.19531	.83660	47
14	.54805	.82465	.65521	.52622	.19553	.83645	46
15	0.54829	1.82384	0.65563	1.52525	1.19576	0.83629	45
16	.54854	.82303	.65604	.52429	.19599	.83613	44
17	.54878	.82222	.65646	.52332	.19622	.83597	43
18	.54902	.82142	.65688	.52235	.19645	.83581	42
19	.54927	.82061	.65729	.52139	.19668	.83565	41
20	0.54951	1.81981	0.65771	1.52043	1.19691	0.83549	40
21	.54975	.81900	.65813	.51946	.19713	.83533	39
22	.54999	.81820	.65854	.51850	.19736	.83517	38
23	.55024	.81740	.65896	.51754	.19759	.83501	37
24	.55048	.81659	.65938	.51658	.19782	.83485	36
25	0.55072	1.81579	0.65980	1.51562	1.19805	0.83469	35
26	.55097	.81499	.66021	.51466	.19828	.83453	34
27	.55121	.81419	.66063	.51370	.19851	.83437	33
28	.55145	.81340	.66105	.51275	.19874	.83421	32
29	.55169	.81260	.66147	.51179	.19897	.83405	31
30	0.55194	1.81180	0.66189	1.51084	1.19920	0.83389	30
31	.55218	.81101	.66230	.50988	.19944	.83373	29
32	.55242	.81021	.66272	.50893	.19967	.83356	28
33	.55266	.80942	.66314	.50797	.19990	.83340	27
34	.55291	.80862	.66356	.50702	.20013	.83324	26
35	0.55315	1.80783	0.66398	1.50607	1.20036	0.83308	25
36	.55339	.80704	.66440	.50512	.20059	.83292	24
37	.55363	.80625	.66482	.50417	.20083	.83276	23
38	.55388	.80546	.66524	.50322	.20106	.83260	22
39	.55412	.80467	.66566	.50228	.20129	.83244	21
40	0.55436	1.80388	0.66608	1.50133	1.20152	0.83228	20
41	.55460	.80309	.66650	.50038	.20176	.83212	19
42	.55484	.80231	.66692	.49944	.20199	.83195	18
43	.55509	.80152	.66734	.49849	.20222	.83179	17
44	.55533	.80074	.66776	.49755	.20246	.83163	16
45	0.55557	1.79995	0.66818	1.49661	1.20269	0.83147	15
46	.55581	.79917	.66860	.49566	.20292	.83131	14
47	.55605	.79839	.66902	.49472	.20316	.83115	13
48	.55630	.79761	.66944	.49378	.20339	.83098	12
49	.55654	.79682	.66986	.49284	.20363	.83082	11
50	0.55678	1.79604	0.67028	1.49190	1.20386	0.83066	10
51	.55702	.79527	.67071	.49097	.20410	.83050	9
52	.55726	.79449	.67113	.49003	.20433	.83034	8
53	.55750	.79371	.67155	.48909	.20457	.83017	7
54	.55775	.79293	.67197	.48816	.20480	.83001	6
55	0.55799	1.79216	0.67239	1.48722	1.20504	0.82985	5
56	.55823	.79138	.67282	.48629	.20527	.82969	4
57	.55847	.79061	.67324	.48536	.20551	.82953	3
58	.55871	.78984	.67366	.48442	.20575	.82936	2
59	.55895	.78906	.67409	.48349	.20598	.82920	1
60	0.55919	1.78829	0.67451	1.48256	1.20622	0.82904	0
123°→ cos	sec	cot	tan	csc	sin ←56°		

34°↓	sin	csc	tan	cot	sec	←145° cos	↓
0	0.55919	1.78829	0.67451	1.48256	1.20622	0.82904	60
1	.55943	.78752	.67493	.48163	.20645	.82887	59
2	.55968	.78675	.67536	.48070	.20669	.82871	58
3	.55992	.78598	.67578	.47977	.20693	.82855	57
4	.56016	.78521	.67620	.47885	.20717	.82839	56
5	0.56040	1.78445	0.67663	1.47792	1.20740	0.82822	55
6	.56064	.78368	.67705	.47699	.20764	.82806	54
7	.56088	.78291	.67748	.47607	.20788	.82790	53
8	.56112	.78215	.67790	.47514	.20812	.82773	52
9	.56136	.78138	.67832	.47422	.20836	.82757	51
10	0.56160	1.78062	0.67875	1.47330	1.20859	0.82741	50
11	.56184	.77986	.67917	.47238	.20883	.82724	49
12	.56208	.77910	.67960	.47146	.20907	.82708	48
13	.56232	.77833	.68002	.47053	.20931	.82692	47
14	.56256	.77757	.68045	.46962	.20955	.82675	46
15	0.56280	1.77681	0.68088	1.46870	1.20979	0.82659	45
16	.56305	.77606	.68130	.46778	.21003	.82643	44
17	.56329	.77530	.68173	.46686	.21027	.82626	43
18	.56353	.77454	.68215	.46595	.21051	.82610	42
19	.56377	.77378	.68258	.46503	.21075	.82593	41
20	0.56401	1.77303	0.68301	1.46411	1.21099	0.82577	40
21	.56425	.77227	.68343	.46320	.21123	.82561	39
22	.56449	.77152	.68386	.46229	.21147	.82544	38
23	.56473	.77077	.68429	.46137	.21171	.82528	37
24	.56497	.77001	.68471	.46046	.21195	.82511	36
25	0.56521	1.76926	0.68514	1.45955	1.21220	0.82495	35
26	.56545	.76851	.68557	.45864	.21244	.82478	34
27	.56569	.76776	.68600	.45773	.21268	.82462	33
28	.56593	.76701	.68642	.45682	.21292	.82446	32
29	.56617	.76626	.68685	.45592	.21316	.82429	31
30	0.56641	1.76552	0.68728	1.45501	1.21341	0.82413	30
31	.56665	.76477	.68771	.45410	.21365	.82396	29
32	.56689	.76402	.68814	.45320	.21389	.82380	28
33	.56713	.76328	.68857	.45229	.21414	.82363	27
34	.56736	.76253	.68900	.45139	.21438	.82347	26
35	0.56760	1.76179	0.68942	1.45049	1.21462	0.82330	25
36	.56784	.76105	.68985	.44958	.21487	.82314	24
37	.56808	.76031	.69028	.44868	.21511	.82297	23
38	.56832	.75956	.69071	.44778	.21535	.82281	22
39	.56856	.75882	.69114	.44688	.21560	.82264	21
40	0.56880	1.75808	0.69157	1.44598	1.21584	0.82248	20
41	.56904	.75734	.69200	.44508	.21609	.82231	19
42	.56928	.75661	.69243	.44418	.21633	.82214	18
43	.56952	.75587	.69286	.44329	.21658	.82198	17
44	.56976	.75513	.69329	.44239	.21682	.82181	16
45	0.57000	1.75440	0.69372	1.44149	1.21707	0.82165	15
46	.57024	.75366	.69416	.44060	.21731	.82148	14
47	.57047	.75293	.69459	.43970	.21756	.82132	13
48	.57071	.75219	.69502	.43881	.21781	.82115	12
49	.57095	.75146	.69545	.43792	.21805	.82098	11
50	0.57119	1.75073	0.69588	1.43703	1.21830	0.82082	10
51	.57143	.75000	.69631	.43614	.21855	.82065	9
52	.57167	.74927	.69675	.43525	.21879	.82048	8
53	.57191	.74854	.69718	.43436	.21904	.82032	7
54	.57215	.74781	.69761	.43347	.21929	.82015	6
55	0.57238	1.74708	0.69804	1.43258	1.21953	0.81999	5
56	.57262	.74635	.69847	.43169	.21978	.81982	4
57	.57286	.74562	.69891	.43080	.22003	.81965	3
58	.57310	.74490	.69934	.42992	.22028	.81949	2
59	.57334	.74417	.69977	.42903	.22053	.81932	1
60	0.57358	1.74345	0.70021	1.42815	1.22077	0.81915	0
124°↑ cos	sec	cot	tan	csc	sin ←55°↑		

35°↓	sin	csc	tan	cot	sec	←144° cos	↓
0	0.57358	1.74345	0.70021	1.42815	1.22077	0.81915	60
1	.57381	.74272	.70064	.42726	.22102	.81899	59
2	.57405	.74200	.70107	.42638	.22127	.81882	58
3	.57429	.74128	.70151	.42550	.22152	.81865	57
4	.57453	.74056	.70194	.42462	.22177	.81848	56
5	0.57477	1.73983	0.70238	1.42374	1.22202	0.81832	55
6	.57501	.73911	.70281	.42286	.22227	.81815	54
7	.57524	.73840	.70325	.42198	.22252	.81798	53
8	.57548	.73768	.70368	.42110	.22277	.81782	52
9	.57572	.73696	.70412	.42022	.22302	.81765	51
10	0.57596	1.73624	0.70455	1.41934	1.22327	0.81748	50
11	.57619	.73552	.70499	.41847	.22352	.81731	49
12	.57643	.73481	.70542	.41759	.22377	.81714	48
13	.57667	.73409	.70586	.41672	.22402	.81698	47
14	.57691	.73338	.70629	.41584	.22428	.81681	46
15	0.57715	1.73267	0.70673	1.41497	1.22453	0.81664	45
16	.57738	.73195	.70717	.41409	.22478	.81647	44
17	.57762	.73124	.70760	.41322	.22503	.81631	43
18	.57786	.73053	.70804	.41235	.22528	.81614	42
19	.57810	.72982	.70848	.41148	.22554	.81597	41
20	0.57833	1.72911	0.70891	1.41061	1.22579	0.81580	40
21	.57857	.72840	.70935	.40974	.22604	.81563	39
22	.57881	.72769	.70979	.40887	.22629	.81546	38
23	.57904	.72698	.71023	.40800	.22655	.81530	37
24	.57928	.72628	.71066	.40714	.22680	.81513	36
25	0.57952	1.72557	0.71110	1.40627	1.22706	0.81496	35
26	.57976	.72487	.71154	.40540	.22731	.81479	34
27	.57999	.72416	.71198	.40454	.22756	.81462	33
28	.58023	.72346	.71242	.40367	.22782	.81445	32
29	.58047	.72275	.71285	.40281	.22807	.81428	31
30	0.58070	1.72205	0.71329	1.40195	1.22833	0.81412	30
31	.58094	.72135	.71373	.40109	.22858	.81395	29
32	.58118	.72065	.71417	.40022	.22884	.81378	28
33	.58141	.71995	.71461	.39936	.22909	.81361	27
34	.58165	.71925	.71505	.39850	.22935	.81344	26
35	0.58189	1.71855	0.71549	1.39764	1.22960	0.81327	25
36	.58212	.71785	.71593	.39679	.22986	.81310	24
37	.58236	.71715	.71637	.39593	.23012	.81293	23
38	.58260	.71646	.71681	.39507	.23037	.81276	22
39	.58283	.71576	.71725	.39421	.23063	.81259	21
40	0.58307	1.71506	0.71769	1.39336	1.23089	0.81242	20
41	.58330	.71437	.71813	.39250	.23114	.81225	19
42	.58354	.71368	.71857	.39165	.23140	.81208	18
43	.58378	.71298	.71901	.39079	.23166	.81191	17
44	.58401	.71229	.71946	.38994	.23192	.81174	16
45	0.58425	1.71160	0.71990	1.38909	1.23217	0.81157	15
46	.58449	.71091	.72034	.38824	.23243	.81140	14
47	.58472	.71022	.72078	.38738	.23269	.81123	13
48	.58496	.70953	.72122	.38653	.23295	.81106	12
49	.58519	.70884	.72167	.38568	.23321	.81089	11
50	0.58543	1.70815	0.72211	1.38484	1.23347	0.81072	10
51	.58567	.70746	.72255	.38399	.23373	.81055	9
52	.58590	.70677	.72299	.38314	.23398	.81038	8
53	.58614	.70609	.72344	.38229	.23424	.81021	7
54	.58637	.70540	.72388	.38145	.23450	.81004	6
55	0.58661	1.70472	0.72432	1.38060	1.23476	0.80987	5
56	.58684	.70403	.72477	.37976	.23502	.80970	4
57	.58708	.70335	.72521	.37891	.23529	.80953	3
58	.58731	.70267	.72565	.37807	.23555	.80936	2
59	.58755	.70198	.72610	.37722	.23581	.80919	1
60	0.58779	1.70130	0.72654	1.37638	1.23607	0.80902	0
125°↑ cos	sec	cot	tan	csc	sin ←54°↑		

36°→	sin	csc	tan	cot	sec	←143° cos	
0	0.58779	1.70130	0.72654	1.37638	1.23607	0.80902	60
1	.58802	.70062	.72699	.37554	.23633	.80885	59
2	.58826	.69994	.72743	.37470	.23659	.80867	58
3	.58849	.69926	.72788	.37386	.23685	.80850	57
4	.58873	.69858	.72832	.37302	.23711	.80833	56
5	0.58896	1.69790	0.72877	1.37218	1.23738	0.80816	55
6	.58920	.69723	.72921	.37134	.23764	.80799	54
7	.58943	.69655	.72966	.37050	.23790	.80782	53
8	.58967	.69587	.73010	.36967	.23816	.80765	52
9	.58990	.69520	.73055	.36883	.23843	.80748	51
10	0.59014	1.69452	0.73100	1.36800	1.23869	0.80730	50
11	.59037	.69385	.73144	.36716	.23895	.80713	49
12	.59061	.69318	.73189	.36633	.23922	.80696	48
13	.59084	.69250	.73234	.36549	.23948	.80679	47
14	.59108	.69183	.73278	.36466	.23975	.80662	46
15	0.59131	1.69116	0.73323	1.36383	1.24001	0.80644	45
16	.59154	.69049	.73368	.36300	.24028	.80627	44
17	.59178	.68982	.73413	.36217	.24054	.80610	43
18	.59201	.68915	.73457	.36134	.24081	.80593	42
19	.59225	.68848	.73502	.36051	.24107	.80576	41
20	0.59248	1.68782	0.73547	1.35968	1.24134	0.80558	40
21	.59272	.68715	.73592	.35885	.24160	.80541	39
22	.59295	.68648	.73637	.35802	.24187	.80524	38
23	.59318	.68582	.73681	.35719	.24213	.80507	37
24	.59342	.68515	.73726	.35637	.24240	.80489	36
25	0.59365	1.68449	0.73771	1.35554	1.24267	0.80472	35
26	.59389	.68382	.73816	.35472	.24293	.80455	34
27	.59412	.68316	.73861	.35389	.24320	.80438	33
28	.59436	.68250	.73906	.35307	.24347	.80420	32
29	.59459	.68183	.73951	.35224	.24373	.80403	31
30	0.59482	1.68117	0.73996	1.35142	1.24400	0.80386	30
31	.59506	.68051	.74041	.35060	.24427	.80368	29
32	.59529	.67985	.74086	.34978	.24454	.80351	28
33	.59552	.67919	.74131	.34896	.24481	.80334	27
34	.59576	.67853	.74176	.34814	.24508	.80316	26
35	0.59599	1.67788	0.74221	1.34732	1.24534	0.80299	25
36	.59622	.67722	.74267	.34650	.24561	.80282	24
37	.59646	.67656	.74312	.34568	.24588	.80264	23
38	.59669	.67591	.74357	.34487	.24615	.80247	22
39	.59693	.67525	.74402	.34405	.24642	.80230	21
40	0.59716	1.67460	0.74447	1.34323	1.24669	0.80212	20
41	.59739	.67394	.74492	.34242	.24696	.80195	19
42	.59763	.67329	.74538	.34160	.24723	.80178	18
43	.59786	.67264	.74583	.34079	.24750	.80160	17
44	.59809	.67198	.74628	.33998	.24777	.80143	16
45	0.59832	1.67133	0.74674	1.33916	1.24804	0.80125	15
46	.59856	.67068	.74719	.33835	.24832	.80108	14
47	.59879	.67003	.74764	.33754	.24859	.80091	13
48	.59902	.66938	.74810	.33673	.24886	.80073	12
49	.59926	.66873	.74855	.33592	.24913	.80056	11
50	0.59949	1.66809	0.74900	1.33511	1.24940	0.80038	10
51	.59972	.66744	.74946	.33430	.24967	.80021	9
52	.59995	.66679	.74991	.33349	.24995	.80003	8
53	.60019	.66615	.75037	.33268	.25022	.79986	7
54	.60042	.66550	.75082	.33187	.25049	.79968	6
55	0.60065	1.66486	0.75128	1.33107	1.25077	0.79951	5
56	.60089	.66421	.75173	.33026	.25104	.79934	4
57	.60112	.66357	.75219	.32946	.25131	.79916	3
58	.60135	.66292	.75264	.32865	.25159	.79899	2
59	.60158	.66228	.75310	.32785	.25186	.79881	1
60	0.60182	1.66164	0.75355	1.32704	1.25214	0.79864	0
126°→ cos	sec	cot	tan	csc	sin ←53°		

37°→	sin	csc	tan	cot	sec	←142° cos	
0	0.60182	1.66164	0.75355	1.32704	1.25214	0.79864	60
1	.60205	.66100	.75401	.32624	.25241	.79846	59
2	.60228	.66036	.75447	.32544	.25269	.79829	58
3	.60251	.65972	.75492	.32464	.25296	.79811	57
4	.60274	.65908	.75538	.32384	.25324	.79793	56
5	0.60298	1.65844	0.75584	1.32304	1.25351	0.79776	55
6	.60321	.65780	.75629	.32224	.25379	.79758	54
7	.60344	.65717	.75675	.32144	.25406	.79741	53
8	.60367	.65653	.75721	.32064	.25434	.79723	52
9	.60390	.65589	.75767	.31984	.25462	.79706	51
10	0.60414	1.65526	0.75812	1.31904	1.25489	0.79688	50
11	.60437	.65462	.75858	.31825	.25517	.79671	49
12	.60460	.65399	.75904	.31745	.25545	.79653	48
13	.60483	.65335	.75950	.31666	.25572	.79635	47
14	.60506	.65272	.75996	.31586	.25600	.79618	46
15	0.60529	1.65209	0.76042	1.31507	1.25628	0.79600	45
16	.60553	.65146	.76088	.31427	.25656	.79583	44
17	.60576	.65083	.76134	.31348	.25683	.79565	43
18	.60599	.65020	.76180	.31269	.25711	.79547	42
19	.60622	.64957	.76226	.31190	.25739	.79530	41
20	0.60645	1.64894	0.76272	1.31110	1.25767	0.79512	40
21	.60668	.64831	.76318	.31031	.25795	.79494	39
22	.60691	.64768	.76364	.30952	.25823	.79477	38
23	.60714	.64705	.76410	.30873	.25851	.79459	37
24	.60738	.64643	.76456	.30795	.25879	.79441	36
25	0.60761	1.64580	0.76502	1.30716	1.25907	0.79424	35
26	.60784	.64518	.76548	.30637	.25935	.79406	34
27	.60807	.64455	.76594	.30558	.25963	.79388	33
28	.60830	.64393	.76640	.30480	.25991	.79371	32
29	.60853	.64330	.76686	.30401	.26019	.79353	31
30	0.60876	1.64268	0.76733	1.30323	1.26047	0.79335	30
31	.60899	.64206	.76779	.30244	.26075	.79318	29
32	.60922	.64144	.76825	.30166	.26104	.79300	28
33	.60945	.64081	.76871	.30087	.26132	.79282	27
34	.60968	.64019	.76918	.30009	.26160	.79264	26
35	0.60991	1.63957	0.76964	1.29931	1.26188	0.79247	25
36	.61015	.63895	.77010	.29853	.26216	.79229	24
37	.61038	.63834	.77057	.29775	.26245	.79211	23
38	.61061	.63772	.77103	.29696	.26273	.79193	22
39	.61084	.63710	.77149	.29618	.26301	.79176	21
40	0.61107	1.63648	0.77196	1.29541	1.26330	0.79158	20
41	.61130	.63587	.77242	.29463	.26358	.79140	19
42	.61153	.63525	.77289	.29385	.26387	.79122	18
43	.61176	.63464	.77335	.29307	.26415	.79105	17
44	.61199	.63402	.77382	.29229	.26443	.79087	16
45	0.61222	1.63341	0.77428	1.29152	1.26472	0.79069	15
46	.61245	.63279	.77475	.29074	.26500	.79051	14
47	.61268	.63218	.77521	.28997	.26529	.79033	13
48	.61291	.63157	.77568	.28919	.26557	.79016	12
49	.61314	.63096	.77615	.28842	.26586	.78998	11
50	0.61337	1.63035	0.77661	1.28764	1.26615	0.78980	10
51	.61360	.62974	.77708	.28687	.26643	.78962	9
52	.61383	.62913	.77754	.28610	.26672	.78944	8
53	.61406	.62852	.77801	.28533	.26701	.78926	7
54	.61429	.62791	.77848	.28456	.26729	.78908	6
55	0.61451	1.62730	0.77895	1.28379	1.26758	0.78891	5
56	.61474	.62669	.77941	.28302	.26787	.78873	4
57	.61497	.62609	.77988	.28225	.26815	.78855	3
58	.61520	.62548	.78035	.28148	.26844	.78837	2
59	.61543	.62487	.78082	.28071	.26873	.78819	1
60	0.61566	1.62427	0.78129	1.27994	1.26902	0.78801	0
127°→ cos	sec	cot	tan	csc	sin ←52°		

38°→	sin	csc	tan	cot	sec	←141° cos	
0	0.61566	1.62427	0.78129	1.27994	1.26902	0.78801	60
1	.61589	.62366	.78175	.27917	.26931	.78783	59
2	.61612	.62306	.78222	.27841	.26960	.78765	58
3	.61635	.62246	.78269	.27764	.26988	.78747	57
4	.61658	.62185	.78316	.27688	.27017	.78729	56
5	0.61681	1.62125	0.78363	1.27611	1.27046	0.78711	55
6	.61704	.62065	.78410	.27535	.27075	.78694	54
7	.61726	.62005	.78457	.27458	.27104	.78676	53
8	.61749	.61945	.78504	.27382	.27133	.78658	52
9	.61772	.61885	.78551	.27306	.27162	.78640	51
10	0.61795	1.61825	0.78598	1.27230	1.27191	0.78622	50
11	.61818	.61765	.78645	.27153	.27221	.78604	49
12	.61841	.61705	.78692	.27077	.27250	.78586	48
13	.61864	.61646	.78739	.27001	.27279	.78568	47
14	.61887	.61586	.78786	.26925	.27308	.78550	46
15	0.61909	1.61526	0.78834	1.26849	1.27337	0.78532	45
16	.61932	.61467	.78881	.26774	.27366	.78514	44
17	.61955	.61407	.78928	.26698	.27396	.78496	43
18	.61978	.61348	.78975	.26622	.27425	.78478	42
19	.62001	.61288	.79022	.26546	.27454	.78460	41
20	0.62024	1.61229	0.79070	1.26471	1.27483	0.78442	40
21	.62046	.61170	.79117	.26395	.27513	.78424	39
22	.62069	.61111	.79164	.26319	.27542	.78405	38
23	.62092	.61051	.79212	.26244	.27572	.78387	37
24	.62115	.60992	.79259	.26169	.27601	.78369	36
25	0.62138	1.60933	0.79306	1.26093	1.27630	0.78351	35
26	.62160	.60874	.79354	.26018	.27660	.78333	34
27	.62183	.60815	.79401	.25943	.27689	.78315	33
28	.62206	.60756	.79449	.25867	.27719	.78297	32
29	.62229	.60698	.79496	.25792	.27748	.78279	31
30	0.62251	1.60639	0.79544	1.25717	1.27778	0.78261	30
31	.62274	.60580	.79591	.25642	.27807	.78243	29
32	.62297	.60521	.79639	.25567	.27837	.78225	28
33	.62320	.60463	.79686	.25492	.27867	.78206	27
34	.62342	.60404	.79734	.25417	.27896	.78188	26
35	0.62365	1.60346	0.79781	1.25343	1.27926	0.78170	25
36	.62388	.60287	.79829	.25268	.27956	.78152	24
37	.62411	.60229	.79877	.25193	.27985	.78134	23
38	.62433	.60171	.79924	.25118	.28015	.78116	22
39	.62456	.60112	.79972	.25044	.28045	.78098	21
40	0.62479	1.60054	0.80020	1.24969	1.28075	0.78079	20
41	.62502	.59996	.80067	.24895	.28105	.78061	19
42	.62524	.59938	.80115	.24820	.28134	.78043	18
43	.62547	.59880	.80163	.24746	.28164	.78025	17
44	.62570	.59822	.80211	.24672	.28194	.78007	16
45	0.62592	1.59764	0.80258	1.24597	1.28224	0.77988	15
46	.62615	.59706	.80306	.24523	.28254	.77970	14
47	.62638	.59648	.80354	.24449	.28284	.77952	13
48	.62660	.59590	.80402	.24375	.28314	.77934	12
49	.62683	.59533	.80450	.24301	.28344	.77916	11
50	0.62706	1.59475	0.80498	1.24227	1.28374	0.77897	10
51	.62728	.59418	.80546	.24153	.28404	.77879	9
52	.62751	.59360	.80594	.24079	.28434	.77861	8
53	.62774	.59302	.80642	.24005	.28464	.77843	7
54	.62796	.59245	.80690	.23931	.28495	.77824	6
55	0.62819	1.59188	0.80738	1.23858	1.28525	0.77806	5
56	.62842	.59130	.80786	.23784	.28555	.77788	4
57	.62864	.59073	.80834	.23710	.28585	.77769	3
58	.62887	.59016	.80882	.23637	.28615	.77751	2
59	.62909	.58959	.80930	.23563	.28646	.77733	1
60	.62932	1.58902	0.80978	1.23490	1.28676	0.77715	0
128°→ cos		sec	cot	tan	csc	sin ←51°	

39°→	sin	csc	tan	cot	sec	←140° cos	
0	0.62932	1.58902	0.80978	1.23490	1.28676	0.77715	60
1	.62955	.58845	.81027	.23416	.28706	.77696	59
2	.62977	.58788	.81075	.23343	.28737	.77678	58
3	.63000	.58731	.81123	.23270	.28767	.77660	57
4	.63022	.58674	.81171	.23196	.28797	.77641	56
5	0.63045	1.58617	0.81220	1.23123	1.28828	0.77623	55
6	.63068	.58560	.81268	.23050	.28858	.77605	54
7	.63090	.58503	.81316	.22977	.28889	.77586	53
8	.63113	.58447	.81364	.22904	.28919	.77568	52
9	.63135	.58390	.81413	.22831	.28950	.77550	51
10	0.63158	1.58333	0.81461	1.22758	1.28980	0.77531	50
11	.63180	.58277	.81510	.22685	.29011	.77513	49
12	.63203	.58221	.81558	.22612	.29042	.77494	48
13	.63225	.58164	.81606	.22539	.29072	.77476	47
14	.63248	.58108	.81655	.22467	.29103	.77458	46
15	0.63271	1.58051	0.81703	1.22394	1.29133	0.77439	45
16	.63293	.57995	.81752	.22321	.29164	.77421	44
17	.63316	.57939	.81800	.22249	.29195	.77402	43
18	.63338	.57883	.81849	.22176	.29226	.77384	42
19	.63361	.57827	.81898	.22104	.29256	.77366	41
20	0.63383	1.57771	0.81946	1.22031	1.29287	0.77347	40
21	.63406	.57715	.81995	.21959	.29318	.77329	39
22	.63428	.57659	.82044	.21886	.29349	.77310	38
23	.63451	.57603	.82092	.21814	.29380	.77292	37
24	.63473	.57547	.82141	.21742	.29411	.77273	36
25	0.63496	1.57491	0.82190	1.21670	1.29442	0.77255	35
26	.63518	.57436	.82238	.21598	.29473	.77236	34
27	.63540	.57380	.82287	.21526	.29504	.77218	33
28	.63563	.57324	.82336	.21454	.29535	.77199	32
29	.63585	.57269	.82385	.21382	.29566	.77181	31
30	0.63608	1.57213	0.82434	1.21310	1.29597	0.77162	30
31	.63630	.57158	.82483	.21238	.29628	.77144	29
32	.63653	.57103	.82531	.21166	.29659	.77125	28
33	.63675	.57047	.82580	.21094	.29690	.77107	27
34	.63698	.56992	.82629	.21023	.29721	.77088	26
35	0.63720	1.56937	0.82678	1.20951	1.29752	0.77070	25
36	.63742	.56881	.82727	.20879	.29784	.77051	24
37	.63765	.56826	.82776	.20808	.29815	.77033	23
38	.63787	.56771	.82825	.20736	.29846	.77014	22
39	.63810	.56716	.82874	.20665	.29877	.76996	21
40	0.63832	1.56661	0.82923	1.20593	1.29909	0.76977	20
41	.63854	.56606	.82972	.20522	.29940	.76959	19
42	.63877	.56551	.83022	.20451	.29971	.76940	18
43	.63899	.56497	.83071	.20379	.30003	.76921	17
44	.63922	.56442	.83120	.20308	.30034	.76903	16
45	0.63944	1.56387	0.83169	1.20237	1.30066	0.76884	15
46	.63966	.56332	.83218	.20166	.30097	.76866	14
47	.63989	.56278	.83268	.20095	.30129	.76847	13
48	.64011	.56223	.83317	.20024	.30160	.76828	12
49	.64033	.56169	.83366	.19953	.30192	.76810	11
50	0.64056	1.56114	0.83415	1.19882	1.30223	0.76791	10
51	.64078	.56060	.83465	.19811	.30255	.76772	9
52	.64100	.56005	.83514	.19740	.30287	.76754	8
53	.64123	.55951	.83564	.19669	.30318	.76735	7
54	.64145	.55897	.83613	.19599	.30350	.76717	6
55	0.64167	1.55843	0.83662	1.19528	1.30382	0.76698	5
56	.64190	.55789	.83712	.19457	.30413	.76679	4
57	.64212	.55734	.83761	.19387	.30445	.76661	3
58	.64234	.55680	.83811	.19316	.30477	.76642	2
59	.64256	.55626	.83860	.19246	.30509	.76623	1
60	0.64279	1.55572	0.83910	1.19175	1.30541	0.76604	0
129°→ cos		sec	cot	tan	csc	sin ←50°	

40°→	sin	csc	tan	cot	sec	cos	←139°
0	0.64279	1.55572	0.83910	1.19175	1.30541	0.76604	60
1	.64301	.55518	.83960	.19105	.30573	.76586	59
2	.64323	.55465	.84009	.19035	.30605	.76567	58
3	.64346	.55411	.84059	.18964	.30636	.76548	57
4	.64368	.55357	.84108	.18894	.30668	.76530	56
5	0.64390	1.55303	0.84158	1.18824	1.30700	0.76511	55
6	.64412	.55250	.84208	.18754	.30732	.76492	54
7	.64435	.55196	.84258	.18684	.30764	.76473	53
8	.64457	.55143	.84307	.18614	.30796	.76455	52
9	.64479	.55089	.84357	.18544	.30829	.76436	51
10	0.64501	1.55036	0.84407	1.18474	1.30861	0.76417	50
11	.64524	.54982	.84457	.18404	.30893	.76398	49
12	.64546	.54929	.84507	.18334	.30925	.76380	48
13	.64568	.54876	.84556	.18264	.30957	.76361	47
14	.64590	.54822	.84606	.18194	.30989	.76342	46
15	0.64612	1.54769	0.84656	1.18125	1.31022	0.76323	45
16	.64635	.54716	.84706	.18055	.31054	.76304	44
17	.64657	.54663	.84756	.17986	.31086	.76286	43
18	.64679	.54610	.84806	.17916	.31119	.76267	42
19	.64701	.54557	.84856	.17846	.31151	.76248	41
20	0.64723	1.54504	0.84906	1.17777	1.31183	0.76229	40
21	.64746	.54451	.84956	.17708	.31216	.76210	39
22	.64768	.54398	.85006	.17638	.31248	.76192	38
23	.64790	.54345	.85057	.17569	.31281	.76173	37
24	.64812	.54292	.85107	.17500	.31313	.76154	36
25	0.64834	1.54240	0.85157	1.17430	1.31346	0.76135	35
26	.64856	.54187	.85207	.17361	.31378	.76116	34
27	.64878	.54134	.85257	.17292	.31411	.76097	33
28	.64901	.54082	.85308	.17223	.31443	.76078	32
29	.64923	.54029	.85358	.17154	.31476	.76059	31
30	0.64945	1.53977	0.85408	1.17085	1.31509	0.76041	30
31	.64967	.53924	.85458	.17016	.31541	.76022	29
32	.64989	.53872	.85509	.16947	.31574	.76003	28
33	.65011	.53820	.85559	.16878	.31607	.75984	27
34	.65033	.53768	.85609	.16809	.31640	.75965	26
35	0.65055	1.53715	0.85660	1.16741	1.31672	0.75946	25
36	.65077	.53663	.85710	.16672	.31705	.75927	24
37	.65100	.53611	.85761	.16603	.31738	.75908	23
38	.65122	.53559	.85811	.16535	.31771	.75889	22
39	.65144	.53507	.85862	.16466	.31804	.75870	21
40	0.65166	1.53455	0.85912	1.16398	1.31837	0.75851	20
41	.65188	.53403	.85963	.16329	.31870	.75832	19
42	.65210	.53351	.86014	.16261	.31903	.75813	18
43	.65232	.53299	.86064	.16192	.31936	.75794	17
44	.65254	.53247	.86115	.16124	.31969	.75775	16
45	0.65276	1.53196	0.86166	1.16056	1.32002	0.75756	15
46	.65298	.53144	.86216	.15987	.32035	.75738	14
47	.65320	.53092	.86267	.15919	.32068	.75719	13
48	.65342	.53041	.86318	.15851	.32101	.75700	12
49	.65364	.52989	.86368	.15783	.32134	.75680	11
50	0.65386	1.52938	0.86419	1.15715	1.32168	0.75661	10
51	.65408	.52886	.86470	.15647	.32201	.75642	9
52	.65430	.52835	.86521	.15579	.32234	.75623	8
53	.65452	.52784	.86572	.15511	.32267	.75604	7
54	.65474	.52732	.86623	.15443	.32301	.75585	6
55	0.65496	1.52681	0.86674	1.15375	1.32334	0.75566	5
56	.65518	.52630	.86725	.15308	.32368	.75547	4
57	.65540	.52579	.86776	.15240	.32401	.75528	3
58	.65562	.52527	.86827	.15172	.32434	.75509	2
59	.65584	.52476	.86878	.15104	.32468	.75490	1
60	0.65606	1.52425	0.86929	1.15037	1.32501	0.75471	0
130°→	cos	sec	cot	tan	csc	sin	←49°

41°→	sin	csc	tan	cot	sec	cos	←138°
0	0.65606	1.52425	0.86929	1.15037	1.32501	0.75471	60
1	.65628	.52374	.86980	.14969	.32535	.75452	59
2	.65650	.52323	.87031	.14902	.32568	.75433	58
3	.65672	.52273	.87082	.14834	.32602	.75414	57
4	.65694	.52222	.87133	.14767	.32636	.75395	56
5	0.65716	1.52171	0.87184	1.14699	1.32669	0.75375	55
6	.65738	.52120	.87236	.14632	.32703	.75356	54
7	.65759	.52069	.87287	.14565	.32737	.75337	53
8	.65781	.52019	.87338	.14498	.32770	.75318	52
9	.65803	.51968	.87389	.14430	.32804	.75299	51
10	0.65825	1.51918	0.87441	1.14363	1.32838	0.75280	50
11	.65847	.51867	.87492	.14296	.32872	.75261	49
12	.65869	.51817	.87543	.14229	.32905	.75241	48
13	.65891	.51766	.87595	.14162	.32939	.75222	47
14	.65913	.51716	.87646	.14095	.32973	.75203	46
15	0.65935	1.51665	0.87698	1.14028	1.33007	0.75184	45
16	.65956	.51615	.87749	.13961	.33041	.75165	44
17	.65978	.51565	.87801	.13894	.33075	.75146	43
18	.66000	.51515	.87852	.13828	.33109	.75126	42
19	.66022	.51465	.87904	.13761	.33143	.75107	41
20	0.66044	1.51415	0.87955	1.13694	1.33177	0.75088	40
21	.66066	.51364	.88007	.13627	.33211	.75069	39
22	.66088	.51314	.88059	.13561	.33245	.75050	38
23	.66109	.51265	.88110	.13494	.33279	.75030	37
24	.66131	.51215	.88162	.13428	.33314	.75011	36
25	0.66153	1.51165	0.88214	1.13361	1.33348	0.74992	35
26	.66175	.51115	.88265	.13295	.33382	.74973	34
27	.66197	.51065	.88317	.13228	.33416	.74953	33
28	.66218	.51015	.88369	.13162	.33451	.74934	32
29	.66240	.50966	.88421	.13096	.33485	.74915	31
30	0.66262	1.50916	0.88473	1.13029	1.33519	0.74896	30
31	.66284	.50866	.88524	.12963	.33554	.74876	29
32	.66306	.50817	.88576	.12897	.33588	.74857	28
33	.66327	.50767	.88628	.12831	.33622	.74838	27
34	.66349	.50718	.88680	.12765	.33657	.74818	26
35	0.66371	1.50669	0.88732	1.12699	1.33691	0.74799	25
36	.66393	.50619	.88784	.12633	.33726	.74780	24
37	.66414	.50570	.88836	.12567	.33760	.74760	23
38	.66436	.50521	.88888	.12501	.33795	.74741	22
39	.66458	.50471	.88940	.12435	.33830	.74722	21
40	0.66480	1.50422	0.88992	1.12369	1.33864	0.74703	20
41	.66501	.50373	.89045	.12303	.33899	.74683	19
42	.66523	.50324	.89097	.12238	.33934	.74664	18
43	.66545	.50275	.89149	.12172	.33968	.74644	17
44	.66566	.50226	.89201	.12106	.34003	.74625	16
45	0.66588	1.50177	0.89253	1.12041	1.34038	0.74606	15
46	.66610	.50128	.89306	.11975	.34073	.74586	14
47	.66632	.50079	.89358	.11909	.34108	.74567	13
48	.66653	.50030	.89410	.11844	.34142	.74548	12
49	.66675	.49981	.89463	.11778	.34177	.74528	11
50	0.66697	1.49933	0.89515	1.11713	1.34212	0.74509	10
51	.66718	.49884	.89567	.11648	.34247	.74489	9
52	.66740	.49835	.89620	.11582	.34282	.74470	8
53	.66762	.49787	.89672	.11517	.34317	.74451	7
54	.66783	.49738	.89725	.11452	.34352	.74431	6
55	0.66805	1.49690	0.89777	1.11387	1.34387	0.74412	5
56	.66827	.49641	.89830	.11321	.34423	.74392	4
57	.66848	.49593	.89883	.11256	.34458	.74373	3
58	.66870	.49544	.89935	.11191	.34493	.74353	2
59	.66891	.49496	.89988	.11126	.34528	.74334	1
60	0.66913	1.49448	0.90040	1.11061	1.34563	0.74314	0
131°→	cos	sec	cot	tan	csc	sin	←48°

42°↓	sin	csc	tan	cot	sec	cos ←137°	
0	0.66913	1.49448	0.90040	1.11061	1.34563	0.74314	60
1	.66935	.49399	.90093	.10996	.34599	.74295	59
2	.66956	.49351	.90146	.10931	.34634	.74276	58
3	.66978	.49303	.90199	.10867	.34669	.74256	57
4	.66999	.49255	.90251	.10802	.34704	.74237	56
5	0.67021	1.49207	0.90304	1.10737	1.34740	0.74217	55
6	.67043	.49159	.90357	.10672	.34775	.74198	54
7	.67064	.49111	.90410	.10607	.34811	.74178	53
8	.67086	.49063	.90463	.10543	.34846	.74159	52
9	.67107	.49015	.90516	.10478	.34882	.74139	51
10	0.67129	1.48967	0.90569	1.10414	1.34917	0.74120	50
11	.67151	.48919	.90621	.10349	.34953	.74100	49
12	.67172	.48871	.90674	.10285	.34988	.74080	48
13	.67194	.48824	.90727	.10220	.35024	.74061	47
14	.67215	.48776	.90781	.10156	.35060	.74041	46
15	0.67237	1.48728	0.90834	1.10091	1.35095	0.74022	45
16	.67258	.48681	.90887	.10027	.35131	.74002	44
17	.67280	.48633	.90940	.09963	.35167	.73983	43
18	.67301	.48586	.90993	.09899	.35203	.73963	42
19	.67323	.48538	.91046	.09834	.35238	.73944	41
20	0.67344	1.48491	0.91099	1.09770	1.35274	0.73924	40
21	.67366	.48443	.91153	.09706	.35310	.73904	39
22	.67387	.48396	.91206	.09642	.35346	.73885	38
23	.67409	.48349	.91259	.09578	.35382	.73865	37
24	.67430	.48301	.91313	.09514	.35418	.73846	36
25	0.67452	1.48254	0.91366	1.09450	1.35454	0.73826	35
26	.67473	.48207	.91419	.09386	.35490	.73806	34
27	.67495	.48160	.91473	.09322	.35526	.73787	33
28	.67516	.48113	.91526	.09258	.35562	.73767	32
29	.67538	.48066	.91580	.09195	.35598	.73747	31
30	0.67559	1.48019	0.91633	1.09131	1.35634	0.73728	30
31	.67580	.47972	.91687	.09067	.35670	.73708	29
32	.67602	.47925	.91740	.09003	.35707	.73688	28
33	.67623	.47878	.91794	.08940	.35743	.73669	27
34	.67645	.47831	.91847	.08876	.35779	.73649	26
35	0.67666	1.47784	0.91901	1.08813	1.35815	0.73629	25
36	.67688	.47738	.91955	.08749	.35852	.73610	24
37	.67709	.47691	.92008	.08686	.35888	.73590	23
38	.67730	.47644	.92062	.08622	.35924	.73570	22
39	.67752	.47598	.92116	.08559	.35961	.73551	21
40	0.67773	1.47551	0.92170	1.08496	1.35997	0.73531	20
41	.67795	.47504	.92224	.08432	.36034	.73511	19
42	.67816	.47458	.92277	.08369	.36070	.73491	18
43	.67837	.47411	.92331	.08306	.36107	.73472	17
44	.67859	.47365	.92385	.08243	.36143	.73452	16
45	0.67880	1.47319	0.92439	1.08179	1.36180	0.73432	15
46	.67901	.47272	.92493	.08116	.36217	.73413	14
47	.67923	.47226	.92547	.08053	.36253	.73393	13
48	.67944	.47180	.92601	.07990	.36290	.73373	12
49	.67965	.47134	.92655	.07927	.36327	.73353	11
50	0.67987	1.47087	0.92709	1.07864	1.36363	0.73333	10
51	.68008	.47041	.92763	.07801	.36400	.73314	9
52	.68029	.46995	.92817	.07738	.36437	.73294	8
53	.68051	.46949	.92872	.07676	.36474	.73274	7
54	.68072	.46903	.92926	.07613	.36511	.73254	6
55	0.68093	1.46857	0.92980	1.07550	1.36548	0.73234	5
56	.68115	.46811	.93034	.07487	.36585	.73215	4
57	.68136	.46765	.93088	.07425	.36622	.73195	3
58	.68157	.46719	.93143	.07362	.36659	.73175	2
59	.68179	.46674	.93197	.07299	.36696	.73155	1
60	0.68200	1.46628	0.93252	1.07237	1.36733	0.73135	0
132°→ cos	sec	cot	tan	csc	sin ←47°		

43°↓	sin	csc	tan	cot	sec	cos ←136°	
0	0.68200	1.46628	0.93252	1.07237	1.36733	0.73135	60
1	.68221	.46582	.93306	.07174	.36770	.73116	59
2	.68242	.46537	.93360	.07112	.36807	.73096	58
3	.68264	.46491	.93415	.07049	.36844	.73076	57
4	.68285	.46445	.93469	.06987	.36881	.73056	56
5	0.68306	1.46400	0.93524	1.06925	1.36919	0.73036	55
6	.68327	.46354	.93578	.06862	.36956	.73016	54
7	.68349	.46309	.93633	.06800	.36993	.72996	53
8	.68370	.46263	.93688	.06738	.37030	.72976	52
9	.68391	.46218	.93742	.06676	.37068	.72957	51
10	0.68412	1.46173	0.93797	1.06613	1.37105	0.72937	50
11	.68434	.46127	.93852	.06551	.37143	.72917	49
12	.68455	.46082	.93906	.06489	.37180	.72897	48
13	.68476	.46037	.93961	.06427	.37218	.72877	47
14	.68497	.45992	.94016	.06365	.37255	.72857	46
15	0.68518	1.45946	0.94071	1.06303	1.37293	0.72837	45
16	.68539	.45901	.94125	.06241	.37330	.72817	44
17	.68561	.45856	.94180	.06179	.37368	.72797	43
18	.68582	.45811	.94235	.06117	.37406	.72777	42
19	.68603	.45766	.94290	.06056	.37443	.72757	41
20	0.68624	1.45721	0.94345	1.05994	1.37481	0.72737	40
21	.68645	.45676	.94400	.05932	.37519	.72717	39
22	.68666	.45631	.94455	.05870	.37556	.72697	38
23	.68688	.45587	.94510	.05809	.37594	.72677	37
24	.68709	.45542	.94565	.05747	.37632	.72657	36
25	0.68730	1.45497	0.94620	1.05685	1.37670	0.72637	35
26	.68751	.45452	.94676	.05624	.37708	.72617	34
27	.68772	.45408	.94731	.05562	.37746	.72597	33
28	.68793	.45363	.94786	.05501	.37784	.72577	32
29	.68814	.45319	.94841	.05439	.37822	.72557	31
30	0.68835	1.45274	0.94896	1.05378	1.37860	0.72537	30
31	.68857	.45229	.94952	.05317	.37898	.72517	29
32	.68878	.45185	.95007	.05255	.37936	.72497	28
33	.68899	.45141	.95062	.05194	.37974	.72477	27
34	.68920	.45096	.95118	.05133	.38012	.72457	26
35	0.68941	1.45052	0.95173	1.05072	1.38051	0.72437	25
36	.68962	.45007	.95229	.05010	.38089	.72417	24
37	.68983	.44963	.95284	.04949	.38127	.72397	23
38	.69004	.44919	.95340	.04888	.38165	.72377	22
39	.69025	.44875	.95395	.04827	.38204	.72357	21
40	0.69046	1.44831	0.95451	1.04766	1.38242	0.72337	20
41	.69067	.44787	.95506	.04705	.38280	.72317	19
42	.69088	.44742	.95562	.04644	.38319	.72297	18
43	.69109	.44698	.95618	.04583	.38357	.72277	17
44	.69130	.44654	.95673	.04522	.38396	.72257	16
45	0.69151	1.44610	0.95729	1.04461	1.38434	0.72236	15
46	.69172	.44567	.95785	.04401	.38473	.72216	14
47	.69193	.44523	.95841	.04340	.38512	.72196	13
48	.69214	.44479	.95897	.04279	.38550	.72176	12
49	.69235	.44435	.95952	.04218	.38589	.72156	11
50	0.69256	1.44391	0.96008	1.04158	1.38628	0.72136	10
51	.69277	.44347	.96064	.04097	.38666	.72116	9
52	.69298	.44304	.96120	.04036	.38705	.72095	8
53	.69319	.44260	.96176	.03976	.38744	.72075	7
54	.69340	.44217	.96232	.03915	.38783	.72055	6
55	0.69361	1.44173	0.96288	1.03855	1.38822	0.72035	5
56	.69382	.44129	.96344	.03794	.38860	.72015	4
57	.69403	.44086	.96400	.03734	.38899	.71995	3
58	.69424	.44042	.96457	.03674	.38938	.71974	2
59	.69445	.43999	.96513	.03613	.38977	.71954	1
60	0.69466	1.43956	0.96569	1.03553	1.39016	0.71934	0
133°→ cos	sec	cot	tan	csc	sin ←46°		

44°→	sin	csc	tan	cot	sec	←135° cos	
′							′
0	0.69466	1.43956	0.96569	1.03553	1.39016	0.71934	60
1	.69487	.43912	.96625	.03493	.39055	.71914	59
2	.69508	.43869	.96681	.03433	.39095	.71894	58
3	.69529	.43826	.96738	.03372	.39134	.71873	57
4	.69549	.43783	.96794	.03312	.39173	.71853	56
5	0.69570	1.43739	0.96850	1.03252	1.39212	0.71833	55
6	.69591	.43696	.96907	.03192	.39251	.71813	54
7	.69612	.43653	.96963	.03132	.39291	.71792	53
8	.69633	.43610	.97020	.03072	.39330	.71772	52
9	.69654	.43567	.97076	.03012	.39369	.71752	51
10	0.69675	1.43524	0.97133	1.02952	1.39409	0.71732	50
11	.69696	.43481	.97189	.02892	.39448	.71711	49
12	.69717	.43438	.97246	.02832	.39487	.71691	48
13	.69737	.43395	.97302	.02772	.39527	.71671	47
14	.69758	.43352	.97359	.02713	.39566	.71650	46
15	0.69779	1.43310	0.97416	1.02653	1.39606	0.71630	45
16	.69800	.43267	.97472	.02593	.39646	.71610	44
17	.69821	.43224	.97529	.02533	.39685	.71590	43
18	.69842	.43181	.97586	.02474	.39725	.71569	42
19	.69862	.43139	.97643	.02414	.39764	.71549	41
20	0.69883	1.43096	0.97700	1.02355	1.39804	0.71529	40
21	.69904	.43053	.97756	.02295	.39844	.71508	39
22	.69925	.43011	.97813	.02236	.39884	.71488	38
23	.69946	.42968	.97870	.02176	.39924	.71468	37
24	.69966	.42926	.97927	.02117	.39963	.71447	36
25	0.69987	1.42883	0.97984	1.02057	1.40003	0.71427	35
26	.70008	.42841	.98041	.01998	.40043	.71407	34
27	.70029	.42799	.98098	.01939	.40083	.71386	33
28	.70049	.42756	.98155	.01879	.40123	.71366	32
29	.70070	.42714	.98213	.01820	.40163	.71345	31
30	0.70091	1.42672	0.98270	1.01761	1.40203	0.71325	30
31	.70112	.42630	.98327	.01702	.40243	.71305	29
32	.70132	.42587	.98384	.01642	.40283	.71284	28
33	.70153	.42545	.98441	.01583	.40324	.71264	27
34	.70174	.42503	.98499	.01524	.40364	.71243	26
35	0.70195	1.42461	0.98556	1.01465	1.40404	0.71223	25
36	.70215	.42419	.98613	.01406	.40444	.71203	24
37	.70236	.42377	.98671	.01347	.40485	.71182	23
38	.70257	.42335	.98728	.01288	.40525	.71162	22
39	.70277	.42293	.98786	.01229	.40565	.71141	21
40	0.70298	1.42251	0.98843	1.01170	1.40606	0.71121	20
41	.70319	.42209	.98901	.01112	.40646	.71100	19
42	.70339	.42168	.98958	.01053	.40687	.71080	18
43	.70360	.42126	.99016	.00994	.40727	.71059	17
44	.70381	.42084	.99073	.00935	.40768	.71039	16
45	0.70401	1.42042	0.99131	1.00876	1.40808	0.71019	15
46	.70422	.42001	.99189	.00818	.40849	.70998	14
47	.70443	.41959	.99247	.00759	.40890	.70978	13
48	.70463	.41918	.99304	.00701	.40930	.70957	12
49	.70484	.41876	.99362	.00642	.40971	.70937	11
50	0.70505	1.41835	0.99420	1.00583	1.41012	0.70916	10
51	.70525	.41793	.99478	.00525	.41053	.70896	9
52	.70546	.41752	.99536	.00467	.41093	.70875	8
53	.70567	.41710	.99594	.00408	.41134	.70855	7
54	.70587	.41669	.99652	.00350	.41175	.70834	6
55	0.70608	1.41627	0.99710	1.00291	1.41216	0.70813	5
56	.70628	.41586	.99768	.00233	.41257	.70793	4
57	.70649	.41545	.99826	.00175	.41298	.70772	3
58	.70670	.41504	.99884	.00116	.41339	.70752	2
59	.70690	.41463	.99942	.00058	.41380	.70731	1
60	0.70711	1.41421	1.00000	1.00000	1.41421	0.70711	0
134°→ cos	sec	cot	tan	csc	sin ←45°		

TABLE OF SQUARES AND SQUARE ROOTS

n	n²	√n	n	n²	√n	n	n²	√n
1	001	1.0000	51	2601	7.1414	101	10201	10.0498
2	004	1.4142	52	2704	7.2111	102	10404	10.0995
3	009	1.7320	53	2809	7.2801	103	10609	10.1488
4	016	2.0000	54	2916	7.3484	104	10816	10.1980
5	025	2.2360	55	3025	7.4161	105	11025	10.2469
6	036	2.4494	56	3136	7.4833	106	11236	10.2956
7	049	2.6457	57	3249	7.5498	107	11449	10.3440
8	064	2.8284	58	3364	7.6157	108	11664	10.3923
9	081	3.0000	59	3481	7.6811	109	11881	10.4403
10	100	3.1622	60	3600	7.7459	110	12100	10.4880
11	121	3.3166	61	3721	7.8102	111	12321	10.5356
12	144	3.4641	62	3844	7.8740	112	12544	10.5830
13	169	3.6055	63	3969	7.9372	113	12769	10.6301
14	196	3.7416	64	4096	8.0000	114	12996	10.6770
15	225	3.8729	65	4225	8.0622	115	13225	10.7238
16	256	4.0000	66	4356	8.1240	116	13456	10.7703
17	289	4.1231	67	4489	8.1853	117	13689	10.8166
18	324	4.2426	68	4624	8.2462	118	13924	10.8627
19	361	4.3589	69	4761	8.3066	119	14161	10.9087
20	400	4.4721	70	4900	8.3666	120	14400	10.9544
21	441	4.5825	71	5041	8.4261	121	14641	11.0000
22	484	4.6904	72	5184	8.4852	122	14884	11.0453
23	529	4.7958	73	5329	8.5440	123	15129	11.0905
24	576	4.8989	74	5476	8.6023	124	15376	11.1355
25	625	5.0000	75	5625	8.6602	125	15625	11.1803
26	676	5.0990	76	5776	8.7178	126	15876	11.2249
27	729	5.1961	77	5929	8.7749	127	16129	11.2694
28	784	5.2915	78	6084	8.8317	128	16384	11.3137
29	841	5.3851	79	6241	8.8882	129	16641	11.3578
30	900	5.4772	80	6400	8.9442	130	16900	11.4017
31	961	5.5677	81	6561	9.0000	131	17161	11.4455
32	1024	5.6568	82	6724	9.0553	132	17424	11.4891
33	1089	5.7445	83	6889	9.1104	133	17689	11.5325
34	1156	5.8309	84	7056	9.1651	134	17956	11.5758
35	1225	5.9160	85	7225	9.2195	135	18225	11.6189
36	1296	6.0000	86	7396	9.2736	136	18496	11.6619
37	1369	6.0827	87	7569	9.3273	137	18769	11.7046
38	1444	6.1644	88	7744	9.3808	138	19044	11.7473
39	1521	6.2449	89	7921	9.4340	139	19321	11.7898
40	1600	6.3245	90	8100	9.4868	140	19600	11.8321
41	1681	6.4031	91	8281	9.5394	141	19881	11.8743
42	1764	6.4807	92	8464	9.5916	142	20164	11.9163
43	1849	6.5574	93	8649	9.6436	143	20449	11.9582
44	1936	6.6332	94	8836	9.6953	144	20736	12.0000
45	2025	6.7082	95	9025	9.7467	145	21025	12.0415
46	2116	6.7823	96	9216	9.7979	146	21316	12.0830
47	2209	6.8556	97	9409	9.8488	147	21609	12.1243
48	2304	6.9282	98	9604	9.8994	148	21904	12.1655
49	2401	7.0000	99	9801	9.9498	149	22201	12.2065
50	2500	7.0710	100	10000	10.0000	150	22500	12.2474

n	n²	√n	n	n²	√n	n	n²	√n
151	22801	12.2882	201	40401	14.1774	251	63001	15.8429
152	23104	12.3288	202	40804	14.2126	252	63504	15.8745
153	23409	12.3693	203	41209	14.2478	253	64009	15.9059
154	23716	12.4096	204	41616	14.2828	254	64516	15.9373
155	24025	12.4498	205	42025	14.3178	255	65025	15.9687
156	24336	12.4899	206	42436	14.3527	256	65536	16.0000
157	24649	12.5299	207	42849	14.3874	257	66049	16.0312
158	24964	12.5698	208	43264	14.4222	258	66564	16.0623
159	25281	12.6095	209	43681	14.4568	259	67081	16.0934
160	25600	12.6491	210	44100	14.4913	260	67600	16.1245
161	25921	12.6885	211	44521	14.5258	261	68121	16.1554
162	26244	12.7279	212	44944	14.5602	262	68644	16.1864
163	26569	12.7671	213	45369	14.5945	263	69169	16.2172
164	26896	12.8062	214	45796	14.6287	264	69696	16.2480
165	27225	12.8452	215	46225	14.6628	265	70225	16.2788
166	27556	12.8840	216	46656	14.6969	266	70756	16.3095
167	27889	12.9228	217	47089	14.7309	267	71289	16.3401
168	28224	12.9614	218	47524	14.7648	268	71824	16.3707
169	28561	13.0000	219	47961	14.7986	269	72361	16.4012
170	28900	13.0384	220	48400	14.8323	270	72900	16.4316
171	29241	13.0766	221	48841	14.8660	271	73441	16.4620
172	29584	13.1148	222	49284	14.8996	272	73984	16.4924
173	29929	13.1529	223	49729	14.9331	273	74529	16.5227
174	30276	13.1909	224	50176	14.9666	274	75076	16.5529
175	30625	13.2287	225	50625	15.0000	275	75625	16.5831
176	30976	13.2664	226	51076	15.0332	276	76176	16.6132
177	31329	13.3041	227	51529	15.0665	277	76729	16.6433
178	31684	13.3416	228	51984	15.0996	278	77284	16.6733
179	32041	13.3790	229	52441	15.1327	279	77841	16.7032
180	32400	13.4164	230	52900	15.1657	280	78400	16.7332
181	32761	13.4536	231	53361	15.1986	281	78961	16.7630
182	33124	13.4907	232	53824	15.2315	282	79524	16.7928
183	33489	13.5277	233	54289	15.2643	283	80089	16.8226
184	33856	13.5646	234	54756	15.2970	284	80656	16.8523
185	34225	13.6014	235	55225	15.3297	285	81225	16.8819
186	34596	13.6381	236	55696	15.3622	286	81796	16.9115
187	34969	13.6747	237	56169	15.3948	287	82369	16.9410
188	35344	13.7113	238	56644	15.4272	288	82944	16.9705
189	35721	13.7477	239	57121	15.4596	289	83521	17.0000
190	36100	13.7840	240	57600	15.4919	290	84100	17.0293
191	36481	13.8202	241	58081	15.5241	291	84681	17.0587
192	36864	13.8564	242	58564	15.5563	292	85264	17.0880
193	37249	13.8924	243	59049	15.5884	293	85849	17.1172
194	37636	13.9283	244	59536	15.6205	294	86436	17.1464
195	38025	13.9642	245	60025	15.6524	295	87025	17.1755
196	38416	14.0000	246	60516	15.6843	296	87616	17.2046
197	38809	14.0356	247	61009	15.7162	297	88209	17.2336
198	39204	14.0712	248	61504	15.7480	298	88804	17.2626
199	39601	14.1067	249	62001	15.7797	299	89401	17.2916
200	40000	14.1421	250	62500	15.8113	300	90000	17.3205

n	n²	√n	n	n²	√n	n	n²	√n
301	90601	17.3493	351	123201	18.7349	401	160801	20.0249
302	91204	17.3781	352	123904	18.7616	402	161604	20.0499
303	91809	17.4069	353	124609	18.7882	403	162409	20.0748
304	92416	17.4356	354	125316	18.8148	404	163216	20.0997
305	93025	17.4642	355	126025	18.8414	405	164025	20.1246
306	93636	17.4928	356	126736	18.8679	406	164836	20.1494
307	94249	17.5214	357	127449	18.8944	407	165649	20.1742
308	94864	17.5499	358	128164	18.9208	408	166464	20.1990
309	95481	17.5784	359	128881	18.9472	409	167281	20.2237
310	96100	17.6068	360	129600	18.9736	410	168100	20.2484
311	96721	17.6352	361	130321	19.0000	411	168921	20.2731
312	97344	17.6635	362	131044	19.0262	412	169744	20.2977
313	97969	17.6918	363	131769	19.0525	413	170569	20.3224
314	98596	17.7200	364	132496	19.0787	414	171396	20.3469
315	99225	17.7482	365	133225	19.1049	415	172225	20.3715
316	99856	17.7764	366	133956	19.1311	416	173056	20.3960
317	100489	17.8045	367	134689	19.1572	417	173889	20.4205
318	101124	17.8325	368	135424	19.1833	418	174724	20.4450
319	101761	17.8605	369	136161	19.2093	419	175561	20.4694
320	102400	17.8885	370	136900	19.2353	420	176400	20.4939
321	103041	17.9164	371	137641	19.2613	421	177241	20.5182
322	103684	17.9443	372	138384	19.2873	422	178084	20.5426
323	104329	17.9722	373	139129	19.3132	423	178929	20.5669
324	104976	18.0000	374	139876	19.3390	424	179776	20.5912
325	105625	18.0277	375	140625	19.3649	425	180625	20.6155
326	106276	18.0554	376	141376	19.3907	426	181476	20.6397
327	106929	18.0831	377	142129	19.4164	427	182329	20.6639
328	107584	18.1107	378	142884	19.4422	428	183184	20.6881
329	108241	18.1383	379	143641	19.4679	429	184041	20.7123
330	108900	18.1659	380	144400	19.4935	430	184900	20.7364
331	109561	18.1934	381	145161	19.5192	431	185761	20.7605
332	110224	18.2208	382	145924	19.5448	432	186624	20.7846
333	110889	18.2482	383	146689	19.5703	433	187489	20.8086
334	111556	18.2756	384	147456	19.5959	434	188356	20.8326
335	112225	18.3030	385	148225	19.6214	435	189225	20.8566
336	112896	18.3303	386	148996	19.6468	436	190096	20.8806
337	113569	18.3575	387	149769	19.6723	437	190969	20.9045
338	114244	18.3847	388	150544	19.6977	438	191844	20.9284
339	114921	18.4119	389	151321	19.7230	439	192721	20.9523
340	115600	18.4390	390	152100	19.7484	440	193600	20.9761
341	116281	18.4661	391	152881	19.7737	441	194481	21.0000
342	116964	18.4932	392	153664	19.7989	442	195364	21.0237
343	117649	18.5202	393	154449	19.8242	443	196249	21.0475
344	118336	18.5472	394	155236	19.8494	444	197136	21.0713
345	119025	18.5741	395	156025	19.8746	445	198025	21.0950
346	119716	18.6010	396	156816	19.8997	446	198916	21.1187
347	120409	18.6279	397	157609	19.9248	447	199809	21.1423
348	121104	18.6547	398	158404	19.9499	448	200704	21.1660
349	121801	18.6815	399	159201	19.9749	449	201601	21.1896
350	122500	18.7082	400	160000	20.0000	450	202500	21.2132

n	n²	√n	n	n²	√n	n	n²	√n
451	203401	21.2367	501	251001	22.3830	551	303601	23.4733
452	204304	21.2602	502	252004	22.4053	552	304704	23.4946
453	205209	21.2837	503	253009	22.4276	553	305809	23.5159
454	206116	21.3072	504	254016	22.4499	554	306916	23.5372
455	207025	21.3307	505	255025	22.4722	555	308025	23.5584
456	207936	21.3541	506	256036	22.4944	556	309136	23.5796
457	208849	21.3775	507	257049	22.5166	557	310249	23.6008
458	209764	21.4009	508	258064	22.5388	558	311364	23.6220
459	210681	21.4242	509	259081	22.5610	559	312481	23.6431
460	211600	21.4476	510	260100	22.5831	560	313600	23.6643
461	212521	21.4709	511	261121	22.6053	561	314721	23.6854
462	213444	21.4941	512	262144	22.6274	562	315844	23.7065
463	214369	21.5174	513	263169	22.6495	563	316969	23.7276
464	215296	21.5406	514	264196	22.6715	564	318096	23.7486
465	216225	21.5638	515	265225	22.6936	565	319225	23.7697
466	217156	21.5870	516	266256	22.7156	566	320356	23.7907
467	218089	21.6101	517	267289	22.7376	567	321489	23.8117
468	219024	21.6333	518	268324	22.7596	568	322624	23.8327
469	219961	21.6564	519	269361	22.7815	569	323761	23.8537
470	220900	21.6794	520	270400	22.8035	570	324900	23.8746
471	221841	21.7025	521	271441	22.8254	571	326041	23.8956
472	222784	21.7255	522	272484	22.8473	572	327184	23.9165
473	223729	21.7485	523	273529	22.8691	573	328329	23.9374
474	224676	21.7715	524	274576	22.8910	574	329476	23.9582
475	225625	21.7944	525	275625	22.9128	575	330625	23.9791
476	226576	21.8174	526	276676	22.9346	576	331776	24.0000
477	227529	21.8403	527	277729	22.9564	577	332929	24.0208
478	228484	21.8632	528	278784	22.9782	578	334084	24.0416
479	229441	21.8860	529	279841	23.0000	579	335241	24.0624
480	230400	21.9089	530	280900	23.0217	580	336400	24.0831
481	231361	21.9317	531	281961	23.0434	581	337561	24.1039
482	232324	21.9544	532	283024	23.0651	582	338724	24.1246
483	233289	21.9772	533	284089	23.0867	583	339889	24.1453
484	234256	22.0000	534	285156	23.1084	584	341056	24.1660
485	235225	22.0227	535	286225	23.1300	585	342225	24.1867
486	236196	22.0454	536	287296	23.1516	586	343396	24.2074
487	237169	22.0680	537	288369	23.1732	587	344569	24.2280
488	238144	22.0907	538	289444	23.1948	588	345744	24.2487
489	239121	22.1133	539	290521	23.2163	589	346921	24.2693
490	240100	22.1359	540	291600	23.2379	590	348100	24.2899
491	241081	22.1585	541	292681	23.2594	591	349281	24.3104
492	242064	22.1810	542	293764	23.2808	592	350464	24.3310
493	243049	22.2036	543	294849	23.3023	593	351649	24.3515
494	244036	22.2261	544	295936	23.3238	594	352836	24.3721
495	245025	22.2485	545	297025	23.3452	595	354025	24.3926
496	246016	22.2710	546	298116	23.3666	596	355216	24.4131
497	247009	22.2934	547	299209	23.3880	597	356409	24.4335
498	248004	22.3159	548	300304	23.4093	598	357604	24.4540
499	249001	22.3383	549	301401	23.4307	599	358801	24.4744
500	250000	22.3606	550	302500	23.4520	600	360000	24.4948

n	n²	√n	n	n²	√n	n	n²	√n
601	361201	24.5153	651	423801	25.5147	701	491401	26.4764
602	362404	24.5356	652	425104	25.5342	702	492804	26.4952
603	363609	24.5560	653	426409	25.5538	703	494209	26.5141
604	364816	24.5764	654	427716	25.5734	704	495616	26.5329
605	366025	24.5967	655	429025	25.5929	705	497025	26.5518
606	367236	24.6170	656	430336	25.6124	706	498436	26.5706
607	368449	24.6373	657	431649	25.6320	707	499849	26.5894
608	369664	24.6576	658	432964	25.6515	708	501264	26.6082
609	370881	24.6779	659	434281	25.6709	709	502681	26.6270
610	372100	24.6981	660	435600	25.6904	710	504100	26.6458
611	373321	24.7184	661	436921	25.7099	711	505521	26.6645
612	374544	24.7386	662	438244	25.7293	712	506944	26.6833
613	375769	24.7588	663	439569	25.7487	713	508369	26.7020
614	376996	24.7790	664	440896	25.7681	714	509796	26.7207
615	378225	24.7991	665	442225	25.7875	715	511225	26.7394
616	379456	24.8193	666	443556	25.8069	716	512656	26.7581
617	380689	24.8394	667	444889	25.8263	717	514089	26.7768
618	381924	24.8596	668	446224	25.8456	718	515524	26.7955
619	383161	24.8797	669	447561	25.8650	719	516961	26.8141
620	384400	24.8997	670	448900	25.8843	720	518400	26.8328
621	385641	24.9198	671	450241	25.9036	721	519841	26.8514
622	386884	24.9399	672	451584	25.9229	722	521284	26.8700
623	388129	24.9599	673	452929	25.9422	723	522729	26.8886
624	389376	24.9799	674	454276	25.9615	724	524176	26.9072
625	390625	25.0000	675	455625	25.9807	725	525625	26.9258
626	391876	25.0199	676	456976	26.0000	726	527076	26.9443
627	393129	25.0399	677	458329	26.0192	727	528529	26.9629
628	394384	25.0599	678	459684	26.0384	728	529984	26.9814
629	395641	25.0798	679	461041	26.0576	729	531441	27.0000
630	396900	25.0998	680	462400	26.0768	730	532900	27.0185
631	398161	25.1197	681	463761	26.0959	731	534361	27.0370
632	399424	25.1396	682	465124	26.1151	732	535824	27.0554
633	400689	25.1594	683	466489	26.1342	733	537289	27.0739
634	401956	25.1793	684	467856	26.1533	734	538756	27.0924
635	403225	25.1992	685	469225	26.1725	735	540225	27.1108
636	404496	25.2190	686	470596	26.1916	736	541696	27.1293
637	405769	25.2388	687	471969	26.2106	737	543169	27.1477
638	407044	25.2586	688	473344	26.2297	738	544644	27.1661
639	408321	25.2784	689	474721	26.2488	739	546121	27.1845
640	409600	25.2982	690	476100	26.2678	740	547600	27.2029
641	410881	25.3179	691	477481	26.2868	741	549081	27.2213
642	412164	25.3377	692	478864	26.3058	742	550564	27.2396
643	413449	25.3574	693	480249	26.3248	743	552049	27.2580
644	414736	25.3771	694	481636	26.3438	744	553536	27.2763
645	416025	25.3968	695	483025	26.3628	745	555025	27.2946
646	417316	25.4165	696	484416	26.3818	746	556516	27.3130
647	418609	25.4361	697	485809	26.4007	747	558009	27.3313
648	419904	25.4558	698	487204	26.4196	748	559504	27.3495
649	421201	25.4754	699	488601	26.4386	749	561001	27.3678
650	422500	25.4950	700	490000	26.4575	750	562500	27.3861

n	n²	√n	n	n²	√n	n	n²	√n
751	564001	27.4043	801	641601	28.3019	851	724201	29.1719
752	565504	27.4226	802	643204	28.3196	852	725904	29.1890
753	567009	27.4408	803	644809	28.3372	853	727609	29.2061
754	568516	27.4590	804	646416	28.3548	854	729316	29.2232
755	570025	27.4772	805	648025	28.3725	855	731025	29.2403
756	571536	27.4954	806	649636	28.3901	856	732736	29.2574
757	573049	27.5136	807	651249	28.4077	857	734449	29.2745
758	574564	27.5318	808	652864	28.4253	858	736164	29.2916
759	576081	27.5499	809	654481	28.4429	859	737881	29.3087
760	577600	27.5680	810	656100	28.4604	860	739600	29.3257
761	579121	27.5862	811	657721	28.4780	861	741321	29.3428
762	580644	27.6043	812	659344	28.4956	862	743044	29.3598
763	582169	27.6224	813	660969	28.5131	863	744769	29.3768
764	583696	27.6405	814	662596	28.5306	864	746496	29.3938
765	585225	27.6586	815	664225	28.5482	865	748225	29.4108
766	586756	27.6767	816	665856	28.5657	866	749956	29.4278
767	588289	27.6947	817	667489	28.5832	867	751689	29.4448
768	589824	27.7128	818	669124	28.6007	868	753424	29.4618
769	591361	27.7308	819	670761	28.6181	869	755161	29.4788
770	592900	27.7488	820	672400	28.6356	870	756900	29.4957
771	594441	27.7668	821	674041	28.6530	871	758641	29.5127
772	595984	27.7848	822	675684	28.6705	872	760384	29.5296
773	597529	27.8028	823	677329	28.6879	873	762129	29.5465
774	599076	27.8208	824	678976	28.7054	874	763876	29.5634
775	600625	27.8388	825	680625	28.7228	875	765625	29.5804
776	602176	27.8567	826	682276	28.7402	876	767376	29.5972
777	603729	27.8747	827	683929	28.7576	877	769129	29.6141
778	605284	27.8926	828	685584	28.7749	878	770884	29.6310
779	606841	27.9105	829	687241	28.7923	879	772641	29.6479
780	608400	27.9284	830	688900	28.8097	880	774400	29.6647
781	609961	27.9463	831	690561	28.8270	881	776161	29.6816
782	611524	27.9642	832	692224	28.8444	882	777924	29.6984
783	613089	27.9821	833	693889	28.8617	883	779689	29.7153
784	614656	28.0000	834	695556	28.8790	884	781456	29.7321
785	616225	28.0178	835	697225	28.8963	885	783225	29.7489
786	617796	28.0356	836	698896	28.9136	886	784996	29.7657
787	619369	28.0535	837	700569	28.9309	887	786769	29.7825
788	620944	28.0713	838	702244	28.9482	888	788544	29.7993
789	622521	28.0891	839	703921	28.9654	889	790321	29.8161
790	624100	28.1069	840	705600	28.9827	890	792100	29.8328
791	625681	28.1247	841	707281	29.0000	891	793881	29.8496
792	627264	28.1424	842	708964	29.0172	892	795664	29.8663
793	628849	28.1602	843	710649	29.0344	893	797449	29.8831
794	630436	28.1780	844	712336	29.0516	894	799236	29.8998
795	632025	28.1957	845	714025	29.0688	895	801025	29.9165
796	633616	28.2134	846	715716	29.0860	896	802816	29.9332
797	635209	28.2311	847	717409	29.1032	897	804609	29.9499
798	636804	28.2488	848	719104	29.1204	898	806404	29.9666
799	638401	28.2665	849	720801	29.1376	899	808201	29.9833
800	640000	28.2842	850	722500	29.1547	900	810000	30.0000

GLOSSARY

ABSOLUTE VALUE — The positive value of a signed number (17)

ADJACENT SIDE — The side which forms the given acute angle with the hypotenuse in a right triangle (33)

ALGEBRAIC EXPRESSION — Contains numbers, signs of operation and letters (variables) (20)

ALGEBRAIC TERMS — A single expression containing combinations of numbers, letters and exponents (20)

ANGLE — An angle is the figure formed by joining two rays at the same point called the origin or vertex (28)

ANGLE MEASURE — Are numbers and can be added together (29)

AUXILLARY LINES — Lines constructed in addition to the given lines for a problem, usually in the form parallel or perpendicular to a given line, to facilitate developing a solution (37)

AXIOM — A rule that is accepted as true without further proof. Also called "postulates" (27)

BASIC DIMENSION — A number used to identify the exact measure or the desired measure between two specific points (15)

BILATERAL TOLERANCE — A number which is added to or subtracted from a basic dimension to establish maximum and minimum limits (15)

BORROWING — The process of changing a mixed number to an equal number having an improper fraction while reducing the whole number by "1". (4)

NOTE:

Numbers in parentheses (   ) following the definition, refer to the unit where a further explanation of the subject can be found.

CANCELLATION — Is used to simplify a fraction to lower terms or to simplify the fractions involved in a multiplication operation (5)

CIRCLE — The set of all points in a plane which are equal distance from a fixed point called the origin (32)

CIRCUMFERENCE — The distance around a circle (32)

COMMON FRACTIONS — A fraction expressed in the form of n/d or numerator followed by a division sign denominator (2)

COMPLEX FRACTION — A fraction with the denominator and/or the numerator expressed as a fraction or mixed number (6)

CROSS MULTIPLICATION — The process used on equations to test for equality (2)

DECIMAL ANGLES — Fractions of degrees expressed as decimals instead of in minutes and seconds (35)

DECIMAL SYSTEM — A set of numbers taken from the general set of fractions whose denominator is a multiple of 10, such as 10, 100, 1000, etc. (7)

DENOMINATOR (BOTTOM NUMBER) — Contains the total number of equal parts in one whole unit. (2)

DIAMETER — A segment which joins two points on the circle and passes through the origin (32)

DIVISION — The simplified operation of subtracting one quantity from another quantity a given number of times.
Form: Dividend $\div$ divisor = quotient + remainder

ELEMENTS OF ALGEBRA — The letters or number and letter combinations expressing algebraic form (2)

ELEMENTS OF ARITHMETIC — The numbers using whole numbers, fractions, decimals and percents to show arithmetic form (2)

EQUAL (EQUIVALENT) FRACTIONS — Two or more fractions having equal value (2)

EQUATION — An expression stating two things are equal (23)

EVALUATE — To determine the numerical value of an algebraic expression (23)

EXPONENT — A number that tells how many times a specific number is used as a factor (18)

FACTOR — A number being multiplied by any other number (s) (5, 18)

FACTORING — A process used in determining the least common denominator in which each denominator is subdivided into prime numbers (3)

FACTORIZATION — A set of two or more factors which are equal to a particular product (18)

FORMULA — A rule written in the form of an equation such as A = S (23)

FRACTION — A number representing a quantity of equal parts taken from a whole unit (2)

GEOMETRY — The study of points, lines and planes in space (27)

HYPOTENUSE — The side opposite the right angle in a right triangle (33)

IMPROPER FRACTION — A common fraction with the numerator equal to or larger than the denominator (2)

INVERSE (RECIPROCAL) — The process where a fraction is turned upside down (6)

INTERPOLATION — The process of calculating the value of a function which lies between two known values (33)

LEAST COMMON DENOMINATOR — The lowest denominator that contains a multiple of every other denominator in an addition or subtraction operation (3)

LIKE FRACTIONS — Fractions having the same denominator (3)

LIKE TERMS — Two or more terms containing the same numbers and with the identical exponents (20)

LIMITS — Numbers used to identify the largest (maximum) and smallest (minimum) measure for a range of measurement (15)

LINE — Contains a set of continuous points which extend infinitely in opposite directions (27)

LOWEST TERMS — A fraction with no common factor remaining in the numerator and denominator. Also called simplest form (2)

MAXIMUM DIMENSION — A number used to identify the largest acceptable measure of a range (15)

MEASURE — A number used to indicate length or distance (15)

METRIC SYSTEM — The most common system of measure in use, uses the meter (equals 39.37 inches) as the standard unit of length. (19)

MINIMUM DIMENSION — A number used to identify the smallest acceptable measure of a range (15)

MIXED NUMBER — A number made of a whole number added to a fraction (2)

MULTIPLICATION — The simplified operation of adding a quantity to another quantity a given number of times.
Form: Factor x Factor(s) = Product

NUMBER LINE — A graphic model of the set of all real or signed numbers used to relate the value of any number to that of zero (17)

NUMERATOR (TOP NUMBER) — Contains a number of equal parts which are being related to the total in a common fraction (2)

OBLIQUE TRIANGLE — A triangle which does not contain a right angle (38)

OPPOSITE SIDE — The side opposite a given angle in a right triangle (33)

PARALLEL — Two or more lines in the same plane that never meet or intersect. Also applies to line segments and rays contained in parallel lines (28)

PERCENT — Percent means per hundred and is a short way of relating a number of parts to a whole unit where the whole unit equals 100 percent (16)

PERPENDICULAR — Two lines that meet to form right angles. Also applies to line segments and rays contained in perpendicular lines (28)

PLACE VALUE — The combining process that requires the grouping of decimals in order to correctly complete the addition or subtraction operations (11)

PLANE — A continuous flat surface extending infinitely (27)

POINT — Is used to identify a specific position or location on a line, in a plane or in space (27)

POWER — A product which contains a given factor and indicated number of times (18)

PRIME NUMBERS — The set of whole numbers which cannot be further subdivided into smaller factors such as 2, 3, 5, 7, etc. (3)

PROPER FRACTION — The common fraction with the numerator less than the denominator and has a value less than one whole unit (2)

PROPOSITION — A rule which has been proven true. Also called "theorems" (27)

QUADRILATERALS — Quadrilaterals are four sided polygons which include the square, the rectangle and the parallelogram (30)

QUANTITY — An expression containing two or more terms such as (2 + 5), (x + 5), etc. (22)

RADIUS — A segment which joins the origin to any point on a circle (32)

RATIO OF SIDES OF AN ANGLE — A fraction used to relate the measures of any two sides of a right triangle (33)

RAY — A set of points on a line extending in only one direction. Also called a half line (28)

RIGHT ANGLE — An angle with measure of exactly 90 degrees (28)

RIGHT TRIANGLE — Any triangle containing a right (90°) angle (30)

ROUNDING OFF — The process of reducing the number of digits in a number to a lesser number of significant digits which will produce a value to the desired degree of accuracy (8)

SEGMENT — The set of any two points on a line and all points between them. Also called a line segment (28)

SHOP FRACTIONS — The set of proper fractions most frequently used in the shop is $\frac{1}{2}$, $\frac{1}{4}$, $\frac{1}{8}$, $\frac{1}{16}$, $\frac{1}{32}$, $\frac{1}{64}$ and multiples of these (2)

SQUARE ROOT — A number that is one of the two equal factors in the factorization of a number (18)

STRAIGHT ANGLE — An angle with measure of exactly 180 degrees (28)

SUBSTITUTION — The process of replacing numbers with numbers of equal value (2)

SUBTRACTION — The operation of determining the difference between two numbers or quantities.
    Form: Minuend − subtrahend = difference

SYMBOLS OF INCLUSION — Symbols that include parentheses ( ), brackets [ ], or braces { }, used to group a set of algebraic factors or terms (21)

TANGENT — A line which intersects a circle at only one point (32)

TERM — A single algebraic expression in the form of 2, x, $2x^2$, etc. (22)

TOLERANCES — Numbers used to determine an acceptable range of measurement (15)

TRIANGLES — Triangles are three sided polygons which include the following types: right, isoscles, equilateral and scalene (30)

TRIGONOMETRY — Triangle measure based on the relationship between the angles and the ratios of the sides (33)

UNILATERAL TOLERANCE — A number which is either added to or subtracted from a basic dimension, but not both (15)

UNIT FRACTIONS — A common fraction with the numerator equal "1" (2)

UNLIKE FRACTIONS — Fractions having unlike or different denominators (3)

UNLIKE TERMS — Two or more terms that differ because they have different letters and/or different exponents (20)

WHOLE NUMBER — A quantity of complete units with no fraction part remainder (2)

INDEX

— A —

Absolute Values   96, 357
Acute Angle   186
Addition Principle of Equality   153
Adjacent Angle   187
Adjacent Side   248, 357
Algebra   127-174
Algebra, Addition & Subtraction   127-133
Algebra, Elements of   19, 358
Algebra, Substituting Numerical
   Values   146-147
Algebra, Symbols & Operations   134-137
Algebraic Equations, Addition
   & Subtraction Principles   153-155
Algebraic Equations, Multiplication
   & Division Principles   157-160
Algebraic Expression   127, 357
Algebraic Form   19
Algebraic Multiplication   139-142
   Quantity by a Quantity   141
   Term by a Quantity   141
   Term by a Term   140
Algebraic Terms   128, 357
Algebraic Terms, Addition of   129
Algebraic Terms, Subtraction of   131
Angle   184, 357
Angle Measures   190, 357
Angle, Ratio of Sides   250-252
Angles, Functions of   250-253
Angles, Types of (By Measure)   186
   Acute   186
   Right   186
   Obtuse   186
   Straight   186
   Reflex   186
   Revolution   186
Angles, Types of (By Relative Position)   187
   Adjacent   187
   Complimentary   187
   Supplementary   187
Angles & Lines   182-187
Angles & Lines, Intersecting   190-194

Angles & Lines Relating to
   Polygons & Circles   201-205
Arc   233
Area of a Circle, Formula   169
Area of a Rectangle, Formula   169
Arithmetic   13-109
Arithmetic, Elements of   19, 358
Arithmetic Form   19
Auxillary Lines   278-279, 357
Axiom   177, 357
Axioms and Propositions   177-181

— B —

Basic Dimension   83, 357
   Bilateral Tolerance   84
   Unilateral Tolerance   84
Bilateral Tolerance   83, 84, 357
Borrowing   34, 357
Braces   135
Brackets   135

— C —

Cancellation   358
Cancellation Law   41
Cartesian Coordinate System   258-259
Center Point   178
Centimeter   113
Central Angle   233
Chord   233
Circle, Area of (formula)   169, 235
Circle, Circumference of (formula)   235
Circles   232-237, 358
Circumference   233, 235, 358
Coefficient   128
Common Fractions   12, 53, 358
Complex Fractions   49, 358
Complimentary Angle   187
Cosecant   257
Cosine   256
Cosines, Law of   315-317
   Forms of   316-317
Cotangent   257

Cross Multiplication   18, 49, 358
   Algebraic Form   18
Cross Products   18

— D —

Decimal Angles   267, 358
Decimal, Fraction Equivalent Chart   64-66
Decimal Fractions   53
Decimal Point, Proper Placement   72
Decimal System   52-78, 358
Decimals, Addition and Subtraction of   67-68
Decimals, Annexation of Zeros   55
Decimals, Change to Percent   89
Decimals, Division of   76-78
Decimals, Multiplication of   72-73
Decimals, Rounding Off   58-60
   General Rule   59
Degrees, Conversion to Minutes
   & Seconds   264
Denominator   12, 358
Diameter   233, 358
Difference   32
Dimensions   83-84
   Basic   83, 84
   Maximum   83
   Minimum   83
Distance Formula   168
Dividend   47, 76
Division   47, 358
Division of Fractions, Algebraic Process   48
Division Principle of Equality   158
Divisor   47, 76

— E —

End Points   178
Equal Fractions   14, 18, 358
Equality, to Determine   160
Equality Principle of Powers   166
Equality Principle of Roots   167
Equations,   146, 358
   Golden Rule   159, 170
   With two or more
     Principles of Equality   158
Equilateral Triangle   202
Equivalent Forms, Examples of   53
Equivalent Fractions   14, 358

Evaluate, to   146, 358
Exponent   106, 128, 359

— F —

Factoring   24, 33, 359
Factorization   106, 359
Factors   39, 105, 359
Form, Simplest   14
Formulas   19, 146, 166-170, 359
Formulas, Solving   168
   Procedure   170
Fraction, Changing to Decimal   61-62
Fraction, Changing to Percent   90
Fraction-Decimal Equivalent Chart   64-66
Fraction-Decimal-Millimeter
   Equivalent Chart   122
Fractions   11-66, 359
   Common   12, 53
   Decimal   53
   Equal   14, 18
   Equivalent   14
   Improper   13, 16
   Proper   12
   Unit   13
Fractions, Addition of   22-26
   Algebraic Form   26
   Like Fractions   22-23
   Mixed Numbers   23
   Unlike Fractions   24
Fractions, Division of   46-49
   Algebraic Form   48
Fractions, Multiplication of   39-42
   Algebraic Form   40
   By a Whole Number   40
   Mixed Number   40
   Simple Fractions   39
   Summary   42
Fractions, Subtraction of   31-34
   Algebraic Form   33
   Mixed Numbers   34
Functions of Angles   250-253
Functions, Values of   261-267
   Calculator Application   264
   Finding   262

— G —

Geometry   176-244

Geometry, Basic Axioms   180-181
Geometry, Basic Terms   178-179
Geometry, Definition of   178, 359
Geometry, Problem Solving   194
Geometry, Propositions   191-193, 204-205,
 216-217, 235-237
Geometry, Summary of Basic
 Axioms & Propositions   238-239
Golden Rule of Equations   159, 170

— H —

Hypotenuse   216, 248, 359

— I —

Improper Fractions   13, 16, 359
Inch/Metric Equivalent Values   114
Inches in Decimals to Millimeters
 (Conversion Table)   124
Index Number   108
Inscribed Angle   234
Interpolation   266, 359
Intersecting Lines   179
Intersecting Lines & Angles   190-200
Intersection Point   178
Inverses   47, 359
Isosceles Triangle   202

— L —

Law of Cosines   315-317
 Forms of   316-317
Law of Sines   306-309
 Forms of   307
Least Common Denominator   24, 25, 359
Letters   128
Lower Limit   84
Lowest Terms   14, 359
Like Fractions   359
 Addition of   22-23
Like Terms   128, 359
Limits   83-84, 359
Line   178-179, 183, 359
Line Segment   183
Line vs. Segment   182
Lines and Angles   182-187
 Intersecting   190-194
 Relating to Polygons and Circles   201-205

— M —

Maximum Dimension (Limit)   83, 360
Measure   83, 360
Meter   113
Metric System   113-124, 360
 Conversion Formula   114
 Converting Dimensions
  without Tolerances   116
 Rounding Toleranced Dimensions   117-120
 Rounding Toleranced Dimensions
  when converting from Metric to
  Inch units   121
Midpoint   178
Millimeter   113
Millimeters to Inches (Conversion Table)   123
Minimum Dimension (Limit)   83, 360
Minuend   32
Mixed Numbers   15, 16, 17, 360
 Addition of   23
 Subtraction of   34
Multiplication   360
Multiplication, Cross   18
Multiplication of Fractions   39-42
 Algebraic Form   40
 Summary   42
Multiplication Principle of Equality   157

— N —

Natural Trigonometric Functions,
 Tables of   327-349
Number Line   95, 360
Number, Raising to a Power   107
Numbers   128
 Prime   24, 25
 Mixed   15, 16, 17
 Whole   15, 362
Numerator   12, 360

— O —

Oblique Triangle   306, 360
Obtuse Angle   186
Opposite Angles   191
Opposite Side   248, 360
Order of Operations Law   136, 147, 170
Origin   184

— P —

Parallel   184, 360
Parallel Lines   179
Parallel Planes   179
Parallelogram   203
Parentheses   135
Percent   88-93, 360
   Change to Decimal   89
   Change to Fraction   90
   Types of Problems   91-93
Perpendicular   184, 360
Perpendicular Lines   179
Perpendicular Planes   179
Place Value   53, 360
Place Value Chart   53
Plane   179, 360
Points   178, 360
Powers   105-109, 361
Principle of Equality   153, 170
Prime Numbers   24, 25, 361
Product   39
Projection of Sides   215-218
Proper Fractions   12, 361
Proposition   177, 361
Propositions, Axioms And   177-181
Pythagorean Theorem   215-218
   Formula for   216

— Q —

Quadrants   255
Quadrilaterals   361
   Basic Types   203
Quantity   139, 361
Quotient   47, 76

— R —

Radical Sign   108
Radicand   108
Radius   232, 361
Range   84
Ratios of Sides of an Angle   361
Ray   184, 361
Reciprocals   47, 359
Rectangle   203
Reflex Angle   186
Related Angles   259-260

Remainder   47, 76
Revolution   186
Right Angle   186, 361
Right Triangle   201, 361
Right Triangle Solutions   271-273
Roots   105-109
Roots & Powers, Summary of Definitions   106
Rounding Off   361
   General Rule   59
Rule for Substitution   15

— S —

Scalene Triangle   202
Secont   234, 257
Sector of a Circle   234
Segment   183, 361
Segment of a Circle   234
Shop Fractions   12, 15, 17, 24, 33, 52, 62, 361
Shop Fractions, Decimal Equivalents   64-65
Simple Fractions, Multiplication of   39
Simplest Form   14
Signed Numbers   95-104
Signed Number Operations, Rules for   97-104
Signs of Operations   128
Sine   256
Sine Bar   280
Sines, Law of   306-309
Square   203
Square Roots   108, 361
Square Root Sign   108
Squares & Square Roots, Tables of   350-355
Steel Rule   17
Straight Angle   186, 361
Substitution   361
   Rule for   15
Subtraction   362
Subtraction of Fractions, Algebraic Form   33
Subtraction Principle of Equality   154
Subtrahend   32
Supplementary Angle   187
Symbols of Inclusion   134, 362
   To Remove   135

— T —

Tangent   234, 256, 362
Tapers   282
   Formula for   147

"a tenth" 54
Term 128, 139, 362
   Lowest 14
Tolerances 82-85, 362
   Bilateral 83
   Unilateral 83
Transversal 192
Triangle 362
Triangles, Basic Types 201-202
   Right 201
   Isosceles 202
   Equilateral 202
   Scalene 202
Trigonometric Functions, Natural,
   Tables of 327-349
   Application 265
Trigonometric Functions,
   Positive & Negative Values 258-259

Trigonometry 247-324
   Definition 362
   Introduction 247-248

— U —

Unilateral Tolerance 83, 84, 362
Unit Circle 255
Unit Fractions 13, 362
Unlike Fractions 362
   Addition of 24
Unlike Terms 129, 362
Upper Limit 84

— V —

Variables 128
Vertex 184
Vertical Angles 191

— W —

Whole Number 15, 362